Portage County: of Place and Time

Portage County: of Place and Time

A Historical Geography of Portage County, Wisconsin

Maurice Perret

Cornerstone Press
Stevens Point, Wisconsin

Printed by Worzalla Publishing.

PORTAGE COUNTY: OF PLACE AND TIME

Published by Cornerstone Press.

International Standard Book Number
0-932310-20-6

Cover Photo: St. Peter's Church
Stevens Point, WI

ACKNOWLEDGEMENTS

Special thanks to Professor Carl O. Sauer, from whom I took an invaluable field study class in geography at the University of California, Berkley in 1941.

Thanks to all students who contributed papers on Portage County, colleagues of the Department of Geography and all those who supplied, volunteered or otherwise contributed information.

Thank you also to Professor Daniel Dieterich and Cornerstone Press. Additional thanks to President Stephanie Raymaker, Editor-in-Chief Thaddeus Brockman, and Business Manager Julie Block.

I hope that this book will show what local geography is.

Contents

Part Four

Physical Geography

Part Five

Recreation and Sport

PREFACE

As I promised the students who took my field study course in geography, I have written a book based on their individual studies so that their work might be useful to people interested in Portage County. This book is not a doctoral dissertation to be read only by specialists, nor is it a novel to be read from start to finish. Instead, it is intended to be a source of information for all those who are interested in the subject. In order to be readable, I have tried to limit scientific terms to those that have no current common equivalent, and to avoid long explanations which are more clearly shown on maps, tables and footnotes. A glossary of uncommon words is found at the end of this text.

As I have already mentioned, the base of this book has been furnished by papers prepared by students who took my field study course in geography at the University of Wisconsin-Stevens Point between 1968 and 1983. To these papers I have also added information from publications and other sources. As it is largely a compilation, I have sometimes used information from the student papers and publications without documentation. However, the papers and publications are both listed in the bibliography. The original copies of the student papers, complete with photographs, maps and tables can be found in the Archives of the Portage County Historical Society at the University of Wisconsin-Stevens Point.

Maurice Perret
Professor Emeritus

HISTORY

After the discovery of the New World, settlements took root in the most adaptable and suitable locations. The powerful nations of Europe claimed parts of North America as their own. The Spanish settled first in the South at St. Augustine, Florida, in 1565. On the East Coast, the English founded their first permanent settlement at Jamestown, Virginia, in 1607. The French explored the interior and claimed all the territory from the Gulf of Mexico to Canada. On the West Coast the Spanish moved toward the center of California, opening missions along the way. The Russians, traveling across Siberia and the Bering Strait, took possession of Alaska and moved along the Southern Coast to California.

Although what is now Portage County was originally part of the French colony, there is no historical evidence that the French settled in Portage County. The nearest French settlements were at Green Bay, a convenient port on Lake Michigan, and at Prairie du Chien, a port on the Mississippi River. The only French influence in Portage County is evident in the name of one stream, the Little Eau Pleine River. In 1763, through the Treaty of Paris, the British obtained control of all land east of the Mississippi, including what is now Portage County. There is no historical evidence that white men traveled through the area. The land was primarily covered with forest and was the hunting and trapping grounds of Indians.

After the Revolution, the region officially became part of the United States, but the British retained control of it until 1812. In 1818 it became part of the territory of Michigan. About that time a French Canadian, John Baptiste DuBay, established a trading post for the American Fur Company on the east bank of the Wisconsin River. It was situated where the river could be forded easily on foot and was, therefore, a strategic Indian crossing from the east to the Black River hunting grounds to the west. DuBay thus became the first white settler in what is now Portage County. Not only was the trading post frequented by Indians and fur traders, but it was a spot familiar to lumbermen who worked the river and to farmers who settled in the region. The site was covered by Lake DuBay when a dam was built in 1942.

The territory of Wisconsin was established by an act of Congress in 1836. That same year a treaty signed with Menominee Indians gave a strip of land called "the Indian strip," three miles (4.8 km) wide along both shores of the Wisconsin River, to the Americans. In 1839 the land was surveyed by Joshua Hathaway, divided into lots and, in 1840, put up for sale by the U.S. Land Office in Mineral Point. In Portage County "the Indian strip" was renamed "the Pinery" because of its rich timber resources. Sawmills were built on the Wisconsin River and on some of its tributaries. Logs were floated down the Wisconsin to the Mississippi River and to St. Louis, an important market for lumber. Both the Southwest and the Great Plains were in great need of lumber to build houses and other structures. There are some indications that even before 1840, sawmills had been built on the Wisconsin River and possibly on Mill Creek. Rapids on the Wisconsin River made floating and navigation difficult. George Stevens, an Illinois lumberman, concluded that a site before the rapids was a suitable location for a warehouse. He built one on the left bank, and this place became known as Stevens Point.

The first Council and House of Representatives of the Wisconsin Territory meeting in Belmont established Portage County. At first the area was small, but in 1841 an act extended its boundaries to make it the third largest county in the Wisconsin Territory which extended east and west across eight townships and north through the central part of the state to the present Wisconsin-Michigan line. In 1846, Columbia County with the town of Portage, was separated from Portage County. Later other southern counties also became new counties.

The village of Plover, located on the trail between lower and upper Wisconsin, was at first named Plover Portage; it became the first county seat when it was detached from the southern counties. Portage County was divided into four precincts: Plover Portage, Grand Rapids, Little Bull Falls and Big Bull Falls. These precincts corresponded to the areas around Plover, Wisconsin Rapids, Mosinee and Wausau. In 1847 there were three more precincts: Stevens Point precinct, DuBay precinct and Eau Claire precinct.

At first the county seat was vaguely indicated as Plover Portage, the area used to bypass the rapids, where a tavern had been built and became the established site of the village. Although Plover had been designated as a county seat in 1841, it was not until 1849-1850 that the courthouse was ready for occupancy, and only in 1857 that the village was incorporated.

Pioneers were lumbermen, attracted to the region by the quality of white pine and the accessibility to water as a means of transporting lumber to markets. With the acquisition of tracts of wooden land, lumbermen with the help of laborers began to log the land. Sawmills were established along the streams. Food and equipment were transported from southern Wisconsin and Illinois. Some laborers, who also came from those areas, recognized that the surrounding soil would be profitable farmland to provide wheat, corn, vegetables and meat for lumber camps. Immigrants from the east also established farms on the open land of the southeastern part of the county. Roads were built to connect logging areas, taverns were opened at crossroads and some of them eventually developed into villages.

In 1849 the county board divided the county into three townships: Plover, Middletown and Bull Falls. Plover occupied the southern part; Bull Falls (which later became the core of Marathon County) occupied the northern part and Middletown occupied the middle section between Plover and what is Mosinee today. In 1850 Marathon County was formed, annexing the northern part of Middletown. The township of Plover was divided into Plover and Grand Rapids.

Although Plover was the county seat, Stevens Point was in a location more favorable to development. It was near the landing on the Wisconsin River from which boats could go as far north as the rapids at Mosinee. Stevens Point became the center of the county attracting businessmen, professional men and craftsmen. The village had been platted in 1847, and it already comprised an area called the Public Square. The federal census of 1850, the first official one for Portage County, gives information about the population at that time. It lists households giving names, sex, age, occupation, if the individual had attended school that year, if the individual was over 20 years and unable to read or write and the value of the real estate property. A special note indicated if the individual was not white. In Portage County the only record of a non-white was a five-year-old mulatto girl, born in Wisconsin. There were no records of Indians, although Jean Baptiste DuBay was of mixed blood as shown on his monument in the Knowlton cemetery:

JEAN BAPTISTE DU BAY
1810 - 1887
SON OF A MENOMINEE INDIAN PRINCESS
SON-IN-LAW OF CHIEF OSHKOSH
TREATY MAKER INTERPRETER
INDIAN TRADER FIRM FRIEND OF WHITE MEN

Either DuBay himself, or the census taker, Anglicized the name to John B. DuBay. Records show that neither he nor his wife were able to read or write, and none of his six children had attended school during the year the census was taken. However, DuBay was a lumberman, and according to a real estate value of $10,000, he was the richest man in town. Another lumberman, J. L. Moore, living in the town of Plover, listed the same wealth in real estate. But in the town of Stevens Point, only 25 persons, including DuBay, listed real estate with a total value of $39,025. Twenty-eight persons in Plover listed real estate for a total of $39,650, and in the town of Grand Rapids only 15 had real estate for a total of $16,225--ranging from $25 to $3,000.

Despite its imperfections, the 1850 federal census is useful. By comparing the following federal census with those taken every ten years until today, the development of the population and changes in its characteristics can be noted.

1850 Portage County census:

	Whites		Free Colored		Total
	Males	Females	Males	Females	
Plover (village)	159	123	---	1	283
Plover (southern/ southeastern)	97	71	---	---	168
Stevens Point	307	151	---	---	458
Grand Rapids	265	76	---	---	341
	828	421	---	1	1,250

1850 census of what is Portage County today:

Plover	256	194	---	---	451
Stevens Point	307	151	---	---	458
	563	345	---	1	909

State or countries of birth of people listed in 1850 census:

	Plover I	Plover II	Stevens Point	Grand Rapids	TOTAL
Wisconsin	53	24	86	52	215
Maine	17	6	16	10	49
New Hampshire	--	2	1	2	5
Vermont	17	16	33	7	73
Massachusetts	--	--	1	1	2

Connecticut	1	7	5	--	13
New York	106	52	127	62	347
New Jersey	--	--	1	--	1
Pennsylvania	14	14	33	30	91
Ohio	16	2	19	12	49
Indiana	--	--	4	7	11
Michigan	4	1	13	4	22
Illinois	10	10	23	16	59
Virginia	2	1	5	1	9
South Carolina	1	--	--	--	1
Georgia	--	--	1	--	1
Kentucky	--	1	1	4	6
Mississippi	--	--	1	--	1
Iowa	1	--	1	1	3
Missouri	1	--	3	1	5
Born/United States	243	136	374	212	965
Canada	21	7	34	83	145
Ireland	10	17	14	13	54
England	2	4	15	8	29
Scotland	--	2	2	1	5
Norway	4	1	1	6	12
Sweden	2	--	--	--	2
Germany	1	--	14	8	23
Born/foreign countries	40	31	80	119	270
Birthplace unknown	--	1	4	10	15

TOTAL 283 nds create high risk conditions for forest fires.□

The age groups of the county indicate a youthful population:

Age	Males	Females	Total
under1 yr.	25	20	43
1-5 yrs.	68	75	143

5-9 yrs.	55	59	114
10-14yrs.	35	31	99
Children	183	185	368
15-19yrs.	61	32	93
20-29yrs.	334	122	456
30-39yrs.	166	60	226
40-49yrs.	53	15	68
50-59yrs.	20	5	25
Adults	634	234	868
60-69yrs.	9	2	11
70-79yrs.	1	-	1
80-89yrs.	1	1	2
Elderlies	11	3	14
Grand total	**828**	**422**	**1250**

The 1850 census lists 17 as the number of persons who died in Portage County during the year ending June 1, 1850--among them nine children under ten, and five between 18 and 23 years. Although child mortality was high, there were families (mostly farmers) with five, six, seven and even ten children.

One man, 63 years of age, lived with his wife and two sons (all born in Vermont). One of the sons had a wife from New York State and three children born in Wisconsin. In a house next door lived another son, also born in Vermont, with a wife from Pennsylvania and two small children born in Wisconsin. An Irishman had a wife and five children born in Vermont and two children born in Wisconsin. Others born in the East, with wives from the respective state, had children born in Pennsylvania, Ohio, Illinois or Wisconsin.

The most numerous age groups (20-29 yrs. and 30-39 yrs.) consisted largely of single men. Some of these men had trades, but most were laborers, often living as groups in the households of lumbermen. Others lived with tavernkeepers in hotels.

Lumbermen and laborers were not permanent settlers. When they had cut the best timber, they moved on. It was believed that farmers who came as pioneers would establish farms that would remain in the family for generations. But this did not happen--very few people listed in the 1850 census left descendants in the area.

There were about 450 family names listed in the townships of Plover and Stevens Point in the 1850 census. However in the 1983 telephone directories for Stevens Point, Almond, Amherst and Rosholt, only about 100 were the same names as those in the 1850 census, and most of those bore no relationship to the pioneers. Thirty-two more names were found in the Wisconsin Rapids directory, but again there were no more than one or two families who had linkage to pioneer ancestry in Plover or Stevens Point.

Recent plat books, especially farms in Almond and in the town of Pine Grove, bear names found in the 1850 census. Examples are Beggs (spelled Baggs in the census), Clark, Johnson, Larson, Precourt and Sanders. Archibald Beggs came from Ireland, married in Vermont, and according to the 1850 census, had seven children living at home. Antoine Precourt immigrated from Canada, married in Maine, and at 29 years, had three children living at home. David Sanders and his wife, both from New York State, had ten children living at home; H.H. Young and his wife were from Maine, with four children living at home. Elisha Larson, age 30, and his wife, age 20, came from Norway. All of the preceding were listed as farmers.

W. Kollock, aged 29, and Francis, aged 17, and probably his sister, came from Canada. In the 1850 census, Kollock was listed as a tavernkeeper with a wife from Illinois. However on the plat map of 1876, this same Kollock was said to be a farmer who owned a piece of land. Today there are still Kollocks with farms in the towns of Pine Grove and Almond. It is probable that some men listed as laborers in the 1850 census became farmers and that their descendants still live on farms in Stevens Point or in nearby villages.

Although it is likely the 1850 census missed some people, it still indicates Portage County as sparsely settled at that time with a total of only 909 inhabitants, compared to the 1980 census which counted 57,420 inhabitants in Portage County. Immigrants came from the east and homesteaded in southeastern, eastern, northeastern and western parts of the county.

Between 1850 and 1860, there was an enormous increase in the population in Portage County, larger than in any other decade thereafter. Census figures increased from 909 to 7,507. During this decade more immigrants came from Europe to the United States than in any previous decade. In the years ending December 31, 1860, 2,968,194 people from foreign lands arrived by ship in the United States. In the previous decade, ending September 30, 1849, there were 1,427,337 immigrants; in the decade ending December 31,1839, there were 538,381 immigrants, and in the decade ending September 30, 1829, there were 128,502 immigrants. Realistically the numbers should be increased by the immigrants who came through Canada and decreased by the number of people who were merchants, and visitors who travelled in and out of the country. Passenger arrivals in the United States have been officially recorded in custom houses since September, 1819.

Some light can be shed on immigration into Portage County from a booklet published in 1857 by Albert G. Ellis and J.G. Tracy, entitled "Stevens Point - Handbook." The book's foreward addresses the public: "The design of this little work is to lay before the reader some of the more prominent features of the country of the upper Wisconsin River, the business which has led to its occupancy and settlement...the capacity for sustaining a population..." Ellis was United States Land Office agent. He wanted to attract settlers, stating that in Portage County, "The northern and northwestern portions are mostly timbered containing considerable quantities of choice government land still in the market--75 cents per acre. The supply of pure timber is said to be inexhaustible."

It was during the decade between 1850 and 1860 that most of the towns in Portage County were organized: Eau Pleine in 1851; Almond and Amherst in 1852; Stockton in 1855; Pine Grove, New Hope, Linwood, Lanark and Belmont in 1856; Hull in 1858 and Sharon in 1859. Stevens Point was incorporated as a city in 1858.

In the 1860 federal census, the following facts about Wisconsin can be found. Wisconsin's population: 772,693 whites; 1,171 free-colored; 613 Indians and 494 half-breeds, making a total of 775,881 inhabitants. Of these, 247,177 were born in Wisconsin. At that time 31,185 people born in Wisconsin moved west and south especially to settle in Minnesota, Iowa, Illinois and California. Of Wisconsin's total census, 498,954 had been born in the United States and 276,927 were foreign-born. Of those born in the United States, about half were born in Wisconsin; 120,000 in New York State; 55,000 in New England and 68,000 in the northern states, east of Wisconsin.

Among the foreign born, about 123,000 were from Germany; 44,000 from Great Britain; 18,000 from British America (Canada and Newfoundland) and 21,500 from Norway. Among occupations, farming was by far the most important with about 94,000 farmers and 31,500 farm laborers. There were 28,000 laborers without their specific occupation named and 1,500 lumbermen. The rest were divided among approximately 280 occupations.

The published book of the 1860 federal census does not give many details on the counties. The population of Portage amounts to 7,507 (4,015 males and 3,488 females) in one table but 7,359 (3,934 males and 3,425 females) in another table. No explanation is given for the discrepancy.

It is possible to form a picture of the county at that time from manuscripts of the census takers. Plover had lost its importance and Stevens Point had now become the central place of the county and the only city. In Plover there were still a hotel, two clergymen, a physician, two lawyers, county treasurer, two saloon keepers, some merchants, craftsmen and laborers. At the same time Stevens Point counted among its citizens: several professional men, merchants, craftsmen and two brewers, both from Germany. There was also a daguerreo artist (a photographer) and an Englishman who gave as an occupation--gentleman.

The first railroad line in Wisconsin, from Milwaukee to Wauwatosa, was built in 1850. Companies began building lines to serve the whole state in order to provide transportation to the east and especially to the west which was considered "the promised land." But in 1860, no railroad line had reached Stevens Point and the river was used as a waterway for boats and for rafting logs. There were stage coaches to various localities and consequently the census numbered among occupations: river pilots, raftsmen and a ferryman who lived in the city.

In the city at that time the number of females was almost equal to the number of males. Among the adult single females were teachers, seamstresses and servants.

Although many Stevens Point residents were natives to Wisconsin, many adults came from New York State or from New England. In many families, the father and the mother came from different states, and in some cases one of them was born in a foreign country. Although there were a number of foreign-born residents, they did not seem to form groups according to their ethnic origins.

Almond and Amherst became small centers with some services; there were stores, a church, a school and saloons. The good land on the outwash plain in the south and southeast were settled by farmers who came primarily from New England, New York, Canada and the British Isles. There were some Irish farmers who settled in the towns of Stockton, Belmont, Almond, Lanark, Plover, Buena Vista and Pine Grove. It does not appear that farmers came in groups from other towns or districts. The census indicates that most farmers moved from their place of birth and married women from other states. It seems that farmers first settled in Pennsylvania, Indiana, Ohio, Michigan and Illinois where they had one or several children. They moved further west, settled, and had other children until they finally came to Wisconsin where their youngest children were born. The immigrants who came later settled in the central and northeastern part of the county where glaciation had taken place and farming was more difficult. There were few flat areas and dense woods. In the northwestern part, settlers occupied land that had been forested. Lumbermen had taken out large trees, leaving settlers the work to clear space for farming.

The towns of Sharon and New Hope are good examples of settlement in areas that were unoccupied before. In Sharon the 1860 census lists 454 inhabitants, of which more than half were born in foreign countries. There were 82 dwellings--in this case, a farm with its buildings and inhabitants of the farmer, his family, relatives and laborers. Sixty-six heads of households were farmers: nine born in America (six in New York State, one each in Maine, Vermont and Pennsylvania), 20 born in Germany (Prussia, Bavaria, Baden, Hanover, Hamburg), 16 from Ireland, six with Irish names from Canada, six from Poland, two from England and two from Norway.

Almost all farmers were married and had children, some born in Wisconsin and others born in one of the states between the coast and Wisconsin. In the case of Matthis Simonis for example, he and his wife were born in Prussia, five children were born in New Jersy, three in Pennsylvania and three in Wisconsin.

Only four people indicated occupations other than farming or as laborers--one lumberman from New York, two carpenters (one from Canada and the other, the son of a farmer and living with the family) and one teacher from Vermont living on a farm. With the exception of the families of American farmers, lumbermen and teachers, all other native Americans were children of foreign immigrants. There were clearly three different groups in which families created bonds through language: Germans, Poles and Irish.

TOPONYMY

When people came to Portage County, they gave names reflecting their ethnic origin to the places and environment around them. The names may be divided into several groups:

1. Names of the natural environment such as lakes, streams, swamps, marshes, ponds, islands, forests and woods.
2. Names of administrative divisions and localities such as hamlets, villages and towns.
3. Names of public buildings as schools, post offices and town halls.
4. Names with religious connotations as churches, chapels, convents and cemeteries.
5. Names of railroad stations and depots.
6. Names of roads, streets, avenues, lanes and courts.

Although Indians were the first people to live in Portage County, little evidence remains that they gave names to places. Some creeks had Indian names, but these names were replaced by others. For example, the Little Eau Pleine River was once known as Girchlahu. The only Indian name that still remains is the Wisconsin which is an adaptation of the original Indian name meaning "river." Tomorrow River is the English translation of the Indian word "waupaca," a name given to the upper part of the stream and, later, to the part of the stream below Amherst. Today Waupaca applies only to the part of the stream outside of Portage County. Originally Waupaca was the name applied to the whole stream. Nevertheless, Waupaca was eventually the name given to Waupaca County.

In recent times Indian names, actually pseudo-Indian, have been given to places. Such as a boy scout camp on Sunset Lake called Chick-A-Gami, no actual origin is determinable. Some sources claim that it is a rendition of "where camping is king." This however, is not the case. At present, no actual translation is known. Mikquano Camp for boys on the shore of Lake Elaine comes from a Menominee word for "turtle." Wazeka or Wazeeka, a lake created in the late 1930s by damming Four Mile Creek in Wood County, was the prize-winning name of a contest and is thought to mean "lake of the land of the pines."

The French did not settle in what is now Portage County but they gave the name "portage" to the city they established near Fort Winnebago. It was here Indians carried (or portaged) their canoes and goods from the Fox River to the Wisconsin River. English and Americans later adopted the term "portage" to designate land, trails or roads over which boats or goods were carried between navigable waters. This is why what is now the Village of Plover was initially named Plover Portage, designating the area between the lower and the upper part of the Wisconsin River which had to be portaged because of a series of rapids. At first it was believed that the portage was between the Wisconsin River and the Waupaca River, but later it was believed Indians crossed the Wisconsin where it was shallow and could easily be crossed on foot. It was at such a place that John Baptiste DuBay established his trading post. The French gave names to the Eau Claire meaning "clear water," and to the Eau Pleine meaning "full water." The Eau Pleine has also been written Aux Pleines River which could mean "river of the plains". The Little Eau Claire and the Little Eau Pleine are branches of the Eau Claire and the Eau Pleine rivers which flow in Marathon County.

Names of natural features sometimes refer to natural conditions, such as Lone Rock, Rock Hill, Standing Rock, The Ledge, Long Island, Mud Lake and Mudhole Lake. Some names reflect a situation: Three Lakes, Twin Lakes and East Lake. Some indicate amenities: Pleasant Lake, Sunset Lake and Silver Lake. Names that refer to trees, fish or wildlife as Pine Lake, Poplar Lake, Tamarack Lake, Bass Lake, Bluebird Lake, Bear Lake, Wolf Lake and Skunk Lake. Some lakes are named after the property owner on the lake or the owner of the entire lake as Adams, Becker, Boelter, Budsberg, Deans, Eberts, Glisezinski, Hintz, Huntleys, Johnson, Kranski, Meyers, Onland, Ostrowski, Oesterle, Pallen, Patterson, Peters, Pierce, Reton, Rinehart and Riley. And some of the names are alterations of the original name: Glisezinski referred to the Glisczinski family but as the combined letters "scz" sounded wrong to an English-speaking cartographer, an "e" was substituted for the "c", and this new spelling was officially adopted. Locally however, the lake was called Jacqueline. Reinhardt was anglicized to its present spelling, Rinehart Lake. Other names given to lakes are the names of individuals: Elaine, Emily, Helen, Jim, Julia, Susan and Thomas.

Some lake names have special origins. Penny Lake was named after a dog owned by a man who operated a cabin resort. Lions Lake was first named Kiolbassa Lake after a man who had a summer home there. It was renamed Lions Lake when the Wisconsin Lions Foundation bought it and established a summer camp for visually handicapped children. In the town of New Hope the Norwegians named the lake not far from the church Praeste Laken (Preacher Lake). It is now Minister Lake. Lake Pacawa was formed when land was excavated in the 1970s to provide fill for the Highway 51 bypass. Though Pacawa might be thought to be an Indian name, actually it is formed from letters in the last names of three owners of property surrounding the man-made lake: Pagel, Calkins and Worzella.

Artificial lakes, mostly called ponds, were named after the towns where they were dug to retain water for mills: Amherst Mill Pond, Jordan Pond, McDill Pond, Rosholt Mill Pond and Springville Pond. Or, ponds were named after the owners of dams used for the owners' businesses, like Bentley Pond and Christensen Pond. Lake DuBay was named in honor of John Baptiste DuBay, who, around 1820, established a trading post near the Wisconsin River. The place has since been covered with water through the creation of an artificial lake. Several lakes are still locally known by the original name given by the first property owner on its shore. Three Lakes designates Tree Lake, Mud Lake and Penny Lake.

Streams can be called rivers when they are fairly large: Wisconsin River, Tomorrow River, Plover River, Little Eau Claire River, Little Eau Pleine River and Little Wolf Rivers--although the last three are small.

Other streams are called creeks, with the exception of Rocky Run in the town of Linwood and Bloody Run in the town of Grant. The latter would have been named after a stream near Detroit. A few creeks owe their names to inhabitants who lived there: Allen, Bradley, Dopp, Emmons, Halladay, Hayden, Murray, Stedman and Stoltenburg. There are also Bear Creek (two of them), Duck Creek, Hay Meadow Creek, Klondike Creek, Love Creek and Mill Creek; and creeks named North, Pearl, Poncho, Rainy and Spring. It is speculated that Flume Creek was named after a Norwegian pioneer, Ole G.Flum (or Flom), who lived in the area. Twomile, Fourmile and Tenmile creeks start in Portage County and flow into Wood County where they join the Wisconsin River. They were not named for their lengths but because they entered the Wisconsin River two, four or ten miles south of the original village of Grand Rapids. Buena Vista Creek was named after the town of Buena Vista. Dewey Marsh and Buena Vista Marsh are named after the towns in which they are located. Fogarty Marsh was named after the owner of the land.

The city of Stevens Point can trace the origin of its name to a pioneer, George Stevens, who came to the area in 1838. He returned the following year to exploit the wealth of white pine stands around Little Bull Falls, now Mosinee. The area had been opened to the settlers by the Indian

Treaty of 1836. It could be reached by water, from Fort Winnebago to the bottom of a series of rapids, then portaged to the beginning of the rapids to Little Bull Falls. It was on a point before the rapids that George Stevens built a shack to store supplies. This spot became known as Stevens Point and eventually became the gateway to the region of pines, now called the Pinery.

Some settlements were named in honor of famous men like General Ulysses Grant who, in 1862, captured Fort Henry. He later became President of the United States. Custer was named after George Armstrong Custer who, in 1876, was killed in the battle of Little Big Horn when his army was defeated by the Indians. The town of Dewey was named after George Dewey, an admiral who was victorious in the battle of Manila Bay, the battle which, in 1898, ended the Spanish power in the Philippines. Garfield's name came from President James Adams Garfield who distinguished himself during the Civil War.

Some places took their names from local people who had some claim to fame. Alban's name came from James S. Alban, the regimental commander during the Civil War who was mortally wounded at Shiloh in 1862. Bancroft was the name of a pioneer Methodist pastor, Warren Gamaliel Bancroft. Ellis got its name from Albert Gallatin Ellis, a pioneer of the Wisconsin Territory who, though born in New York State, came west. He first settled in Green Bay where, in 1837, he was made surveyor-general. In 1853 Ellis became receiver at the government land office in Stevens Point, and became the most prominent man of his time in Portage County. Jerome Nelson, who, in 1855 or 1856, built a dam on the Tomorrow River to provide energy for a sawmill, gave his name to Nelsonville. Thomas H. and Alexander S. McDill, in 1873, platted the village that took their name--McDill. Rosholt was named after John Gilbert Rosholt of Waupaca County, who established a sawmill on the south branch of the Little Wolf River. Whiting took its name from George A. Whiting, founder of the Whiting-Plover Paper Company, and Coddington from W. B.Coddington who platted a village first known as Pine Island.

Other settlements were named after an individual or family who settled there, as Arnott, Carson, Fancher, Heffron, Hull, Keene, Kellner and Meehan. Immigrants brought the names of their birthplaces with them, such as Belmont and Almond, two villages in Allegany County, New York State; Amherst, a town in Nova Scotia; Lanark, in Scotland and Torun, in Poland. Polonia bears evidence of the origin of its settlers.

Settlers have often given names embodying their ideals to places. There are at least 50 Edens in the United States. Besides New Hope in Portage County, there are at least 20 other places of like name in this country. The Bible is another source of town names. Portage County has no Bethel, though there are at least 40 in the country; also, no Bethlehem nor Hebron. But there is the town of Sharon which is probably named after the fertile plain on the coast of Palestine, and the former village of Jordan was most likely named after the stream in Palestine. Casimir was named after the Polish saint, to whom the local church was dedicated.

Plover found its name in a bird common to the area in earlier times. Junction City (originally Junction) and Amherst Junction are both at the junction of two railroads. There was a prominent pine grove in the central part of the town of Pine Grove. Buena Vista, the only Spanish name in the county, was not given by Spanish-speaking people, but was the name given to a tavern in honor of the American victory over the Mexicans in 1847 at Buena Vista in Mexico. There are at least 16 places with the name of Buena Vista in this country. In Portage County, the name of Buena Vista was given to the town, the creek and to the marsh in that area. The origin of the names of Stockton and Blaine are not clear. They may have been individuals or families living in the area, or names of villages from where they came (there are 18 Stocktons and 18 Blaines in the country). Or the towns could have been named for Robert F. Stockton, a naval officer before the Civil War and James G. Blaine, a member of Congress from Maine and unsuccessful candidate for president in 1884. Park Ridge, one of the latest names given to a locality, received its name from inhabitants who requested incorporation of the village in 1938. Like many names given by developers to subdivisions, Park Ridge suggests a pleasant neighborhood.

Since a post office opened in Plover in 1845, 47 post offices have been established in the county. Most post offices were named after the town, village or community in which they were located. Some names, however, were designated by the postmaster who chose the name of a noted person or prominent family in the area. In some cases, it is not clear if the post office took its name from the community or if the community took its name from the post office.

Schools in Stevens Point, Whiting, Plover and Junction City have been named after presidents of the United States: Grant, Jackson, Jefferson, J.F. Kennedy, Madison, McKinley, Roosevelt and Washington. Others are named after noted people in history: Ben Franklin or Emerson. P.J.

Jacobs was named after a leading community businessman. SPASH is the acronym for Stevens Point Area Senior High. Parochial schools bear the name of the parish.

All Catholic churches bear the name of their patrons--either a saint or a religious symbol. Several are names of persons mentioned in the Bible: Mary, the Blessed Mother of Jesus; St. John the Baptist, precursor of Christ; James, Peter, John and Bartholomew--some of the Twelve Apostles; Paul, first deacon who considered himself on a par with the Apostles; Joseph, the husband of the Virgin Mary; saints of early times like Michael, who lived in the ninth century in Bithynia and was founder of a monastic order, or Martin, a pope of the seventh century who was deported to Crimea and died a martyr (another, Saint Martin was Bishop of Tours, France in the fourth century and became the patron of France).

Names can also be attributed to Patrick, who in the fifth century was the apostle of Ireland; Stephen, born a pagan in 970, was baptized in early youth and became king of Hungary and Adalbert, a Czech who lived in the tenth century and became a missionary in Poland and the first Czech bishop of Prague. Casimir, King of Poland, was beatified and became patron of Poland and Lithuania and Stanislaus, who died a martyr in Cracow in 1079 and is patron saint of Poland. Bronislawa (or Bronislava) was a Premontratensian nun and in Cracow was always recognized as a saint. Considered the patroness of Christian charity, Bronislawa lived in the twelfth century. The church in Custer is dedicated to the Immaculate Conception; the church of Polonia is dedicated to the Sacred Heart and the church of Almond is dedicated to the Guardian Angel.

Episcopal churches have been dedicated as well. In Stevens Point, there is the Episcopal Church of the Intercession and in Amherst, St. Olaf Episcopal Church. Olaf was King Olaf II of Norway. In 1028 he fled the country and was exiled to Russia where he was later killed in battle. Olaf is the patron saint of Norway, but is also venerated in other Scandinavian countries and in England.

Lutheran churches too, are dedicated to saints or religious symbols: St. Paul, St. John, Divine Word, Faith, Good Shepherd, Grace, Our Saviour and Trinity. In Stevens Point, or in towns or villages where there is only one church of the denomination, the church is often referred to by the name of the congregation, such as Lutheran Church in Nelsonville, New Hope and Kellner; the Baptist Church in Almond and Bancroft; United Methodist Church in Almond and Amherst; Church of Christ, Assembly of God in Stevens Point; Seventh Day Adventist in Almond and

Whiting; Apostolic Bible Chapel in Plover and Moravian Church in Kellner.

Some congregations have special names for their places of meeting: Beth Israel Temple for the Jewish synagogue and Kingdom Hall for the Jehovah's Witnesses. Other churches simply have the name of the locality in which they stand: Badger Church, Buena Vista Church, North New Hope Church, South New Hope Church and Meehan Community Church.

Some Catholic and Lutheran cemeteries have the name of the accompanying parishes. Others may be called after the towns in which they are located. Some refer to natural features: Forest Cemetery, Elmwood, Lone Pine, Hillside and Oak Grove. Several cemeteries are named after a conspicuous monument, usually to a prominent family of the vicinity: Bickel Cemetery, Holtz, Kimball, Lombard, Maine, Morrill, Pratt and Runkle. According to tradition, Spiritland Cemetery was named because Dewit McLaughlin came to his wife's grave to meet with her spirit.

When railroads were established, depots and stations were designed in areas where passengers and freight might be expected to be heaviest. Stations were named after the city, village or town, as Stocton; a local person, Coddington; or a natural feature, Rocky Run. Amherst Junction and Juntion City owe their names to the fact that two railroad lines met at these points.

Stevens Point, all villages (even some that were never incorporated) and all additions in Portage County have been platted and named. Today it is the practice to number federal and state highways, assign a letter to county roads and name other thoroughfares.

Names are usually given at random, and present a great variety. Most localities have a Main Street. Most street names can be attributed to locally prominent persons or to people who lived or owned property in the area. In Stevens Point, some of the street names are Atwell, Belke, Bliss, Boyington, Brawley, Briggs, Bukolt, Bush, Chase, Clark, Clayton, Clifford, Cornell, Crosby, Cross, Delaney, Dixon, Ellis, Frederick, Gates, Gilkay, Glennon, Green, Heffron, Howard, Isadore, Janick, Joerns, Karner, Koch, Lee, Loyd, Mason, McCulloch, McDonald, Miller, Nebel, Park, Patch, Peck, Phillips, Prais, Prentice, Rice, Rogers, Scott, Shaurette, Simonis, Sims, Smith, Sommers, Stanley, Strong, Thompson, Vincent, Wadleigh, Walker, Warner, Welsby, Whiting, Whitney, Williams, Wilson, Wood, Wyatt and Zinda.

In Whiting, streets are named: Babcock, Bremmer, Conant, Crueger, Currier, Dunbar, Jacobson, McDill, Mead, Morse, Obrecht, Sherman, Wallace and Walter.

Names that Plover streets carry are: Feltz and Porter; in Rosholt are Frederickson and Hanson; in Junction City there are Brown, Case, Gingle and Morgan and Amherst Junction has Nelson.

Some of the town road names are: Anderson, Baker, Barth, Bentley, Bestul, Brandt, Buelow, Clinton, Corrigan, Damrau, Dulske, Eckels, Firkus, Fletcher, Floistad, Grayson, Griffith, Guth, Hamilton, Harvey, Haviland, Hilmer, Hintz, Hotvedt, Johnson, Krause, Krogwold, Lalk, Loberg, Lucht, Lutz, Madley, Maes, Mead, Meyer, Muelde, Otto, Pacyna, Platta, Precourt, Roegge, Saumer, Sherman, Shirek, Stashek, Steward, Stub, Thilo, Walkush, Walter, Ward, Wells and Wiczek.

Given names may be attached to streets. Often they are of the developer and his family. For example, in Stevens Point: Alice, Dan, Frances, Helen, Henrietta, Howard, John, Lora, Lorraine, Martha, Mary, Mary Ann, Matilda, Pearl, Robert and Stanley (English for the Polish Stanislaus). In Whiting, some streets with given names are: Albert, Julie, Lynda and Tommy; in the town of Hull: Barbara, Carol, Gerald, Mary Jane, May, Robert and Virginia; in the village of Plover: Katherine, Larry, Monica and Patricia.

In the town of Plover, north-south avenues bear the names of presidents, in the following order west to east: Hayes, Monroe, Coolidge, Grant, Buchanan, Jackson, Cleveland, Taft, Lincoln, Hoover, Truman, Harding, Eisenhower and Kennedy. Wilson, Madison and Washington are streets parallel to the avenues in Plover.

Stevens Point has honored Cleveland, Grant, Jefferson, Lincoln, Madison and Monroe by naming streets after them. Whiting has Polk; the town of Hull has Kennedy.

Few national figures have streets named after them, but in Stevens Point there is Franklin and Lindbergh, the first man to fly solo across the Atlantic in 1927. Others are Fremont, a popular explorer of the American West; Wayne after General Anthony Wayne, soldier of the American Revolution and an Indian fighter and Pulaski after General Casimir Pulaski, a Polish volunteer in the American Revolutionary War. Pulaski was killed in action in 1779 near Savannah, Georgia.

Some streets are named according to location: Center, Front (in front of the Wisconsin River), Division, North Point, East, West and South streets or drives.

Other streets reflect the names of nearby buildings. In Stevens Point, Academy Street referred to the academy for Catholic girls of St. Joseph's Convent. Church Street designated the place where the first church in Stevens Point was built. College Avenue is on either side of the University of Wisconsin-Stevens Point, formerly the State Teachers College. Depot Street leads to the railroad depot in Stevens Point. Sanatorium Street was named for the nearby sanatorium, now River Pines Living Center. Others are School Street, Post Road and Townhall Road.

Some streets are named after states, as: California, Illinois, Indiana, Michigan, Minnesota, Texas, Vermont and Wisconsin.

Physical characteristics are the origin of a few street names: Ridge Road, River Road, Maple Ridge and Maple Bluff.

Vegetation has often been used in naming streets, in Stevens Point there are Birchwood, Cedar, Cypress, Forest, Maple, Meadow, Oak, Prairie, Pine, Spruce and Tamarack. In Whiting: Acorn, Beech, Birch, Cedar, Cherry, Chestnut, Hickory, Linden, Locust, Pine, Rose and Willow. In Plover: Butternut, Cedar, Chestnut, Elm, Forest, Hickory, Holly, Magnolia, Maple, Meadow, Norway Pine, Oak, Pine, Red Oak, Red Pine, Whispering Pines, White Pine, Willow and Wintergreen.

Few animal names are found in street names but in Stevens Point there is Elk Street, and Plover has several Deer roads.

Names attached to streets by developers and real estate agents to attract buyers are many: Birchwood, Brentwood, Edgewood, Mapleridge, Pleasant, Riverview and Sunset--all in Stevens Point. In Plover, there are Bel Air, Opportunity, Sandy Acres, Sunrise, Sunset and Whispering Pine. In the town of Hull: Conifer, Edgewood, Elmwood, Fairview, Greenview, Greenwood, Highland, Maple Bluff and Ridgewood.

When Park Ridge was incorporated, it held an assortment of names; some were a continuation of the streets of Stevens Point, others referred to family names. In due time, the village took on the task of renaming the streets. North-south were designated as avenues and named in order: Pinecrest, Sunrise, Fieldcrest, Linwood, Greenbriar and Sunset. East-west stretches were called drives, named in order: Park Ridge (the continuation of Main Street in Stevens Point), Crestwood, Ferndale, Woodlawn, Hillcrest, Oakcrest, Maplewood, Hazelwood and Ridgewood.

Nationally, city planners adopted a system of dividing areas in a grid with thoroughfares running north-south and east-west. The exception to this approach however, was when a city, town or village was built on the bank of a stream or lake or along a railroad line, in which case the thoroughfare would parallel or run at right angles with the stream or railroad tracks. The street along the water or the track would often be called Front Street and the parallel streets numbered: First Street, Second Street, etc. The thoroughfares at right angles were called avenues. In grid formation, north-south were designated streets and east-west, avenues. The center street was Main Street; the center avenue might be Broadway, then First Street W, or First Street NW, etc. The same progression held for avenues.

Often streets and avenues were not laid out according to plan, nor did they have names. This is true in Portage County. In Stevens Point, Main Street starts at the Wisconsin River and runs west-east, First Street runs north, as does Second and Third streets. Front Street, limited to a few blocks, is the western limit of the city. Parallel to Main Street are Fourth, Fifth and Sixth avenues. South of Main Street, there are no numbered streets or avenues.

The town of Plover has First and Second streets running east-west, while the village of Plover has First through Ninth streets and Ninth Court running north-south, and Second Avenue running north-south.

Bancroft has First Street North, Second Street North and Third Street North, First Street South and Second Street South--the dividing line being Main Street.

Almond town and village also have numbered streets and avenues. In the town of Almond streets run north-south, First Street starting at the western townline and the other streets being the lines between sections. The avenues start from the south, the first is County Highway D and the following ones are the section lines north of it. However, in the village of Almond First Street corresponds to County Highway A and Second, Third and Fourth streets are east of it at one block intervals.

Amherst Junction was methodically platted. Main Street remains the southern edge of the built-up part of the village. North of Main are First, Second, Third and Fourth streets. Amherst has First and Second avenues. Nelsonville has First and Third streets. Rosholt has First, Second, Fourth and Fifth streets. Junction City has First and Second streets, the latter is interrupted and the western part is called W. Second Street. In the middle of the town of Grant, farmland and wooded areas have signs which follow the section lines, indicating 100th, 110th and 130th streets. There is a 125th Street and, near Kellner, 83rd, 84th, 85th, 86th and 90th streets. They have been named as part of a future extension plan of the city of Wisconsin Rapids.

A number of thoroughfare names have a unique origin. For example, Akron Avenue on the Waushara county line owes its name to a plan made by Adams, Waushara and counties south, to have all east-west thoroughfares in one township with a name beginning with the same letter. Names were then to follow in alphabetical order from north to south. The roads following townlines are called avenues; the ones parallel or diagonal are called drives, lanes or courts. Thus, south of Portage County in Township 20 North are: Akron Court and Akron Lane, Alp Avenue and Alp Drive, Aniwa Lane and Aniwa Drive, Apache Avenue and Apache Drive, Apache Lane and Apache Court, Archer Court and County Trunk O (Badger Avenue), followed by Badger Lane, Beaver Avenue, Beaver Court, Beaver Drive and Beaver Road.

A number of street names can be explained: Soo Marie Street was named after the Soo Line Railroad--originally the line was the Chicago-Minneapolis-St. Paul-Sault Ste. Marie Company. Wilshire Boulevard, the only boulevard in Stevens Point, got its name from a Wilshire Boulevard apparel shop on Main Street. The shop in turn, had taken its name from Wilshire Boulevard, a fashionable avenue in Beverly Hills (Los Angeles). IGA Avenue stands for Independent Grocers Alliance, a nearby supermarket. Frontenac Street may have been named after Chateau Frotenac, a luxurious hotel in Quebec.

Most of the area public parks were named in honor of distinguished citizens or those persons who had contributed money to further recreation. Such as: Atwell, Bukolt, Cate, Consolidated (for Consolidated Papers Incorporated), Frost, Goerke, Iverson, Mead, Peterson, Slomann, Stedman and Tech. In addition, some are named after physical features: Lake Emily, Lake Helen, Lake DuBay, Collins Lake and Standing Rock. Pioneer Park explains its name. Jordan was named after the village of Jordan which no longer exists.

As there is no official record regarding the naming of various places in Portage County, local people, books and newspapers become the only sources. In many cases the origin indicated may be just a guess and some may be completely wrong, although the explanation is clear: some names have been changed to anglicize them or to give them a sense. Clark Street was certainly named after William H. Clarke, justice of the peace, but the final "e" was thought to be a mistake and was eventually dropped. As mentioned before, Lake Glisczinski became Glisezinski, Allay Street became Alley Street, Annes Drives became Anne's Drive and Runkel became Runkle. It is possible that Airline Road in Whiting and Plover was originally Urline Road since an A.W. Uline (or in some documents Urline) was listed in the 1847 Plover census. That same year county commissioners accepted a report from Adam Urline and James Alban for a road beginning at the Wisconsin River which was to run east past Plover to the current range line between today's Plover and Stockton townships. A.W. Urline settled in the eastern part of the county, and is credited with giving the name of his native town (Amherst, Nova Scotia) to the town of Amherst in Portage County, which was incorporated in 1852. He seems however, to have lived in the southern part of the town which in 1856 became Lanark Town. He and probably a son, Leonese J. Urline, were listed in the earliest taxroll of Lanark, dated 1861. As there are no Urlines in Plover, it is likely that the name was modified to Airline. A few names were anglicized in publications, but kept their original form on official documents: Baker Lake instead of Becker, Budesburg Lake instead of Budsberg.

The origin of many names remains uncertain. Is Georgia Street named after the state of Georgia, or a girl? Is Madison Street named after a president, the capital of Wisconsin or a local family? Are the names given to roads, avenues, streets, lanes, courts and drives those of real or imagined people?

Let it be said some names are of real people, some are not. Certainly, imagination can go far: Jersey Street in Stevens Point was named after a herd of Jersey cows on a farm where Jersey Street was established. No explanation has been found for Bingo Lake or Kozy Street. Wake Island Drive in Amherst probably owes its name to a resident who was in the armed forces on Wake Island during World War II. Are Clar-Re Court, Clar-Re Lake, Clar-Re Lane and Jay-Mar Road the combined first part of given names of spouses?

Are the origins of Forest, Green, Park and Wood natural features or families? In Stevens Point, Forest Street is named after Forest Grant, Green Avenue after Mrs. Theresa Green and Park Street after Judge Byron B. Park.

One may hope that some day a scholar will study the documents, interview persons on the origin of names and write a book on "Place Names of Portage County."

SETTLEMENTS: THE CITY, VILLAGES AND FARMS

Stevens Point was platted as a village in 1847. Counting approximately 2,000 residents in 1858, it was incorporated as a city on May 1 of that year. Most of the services needed in the area were in place. In a little more than 100 years the city developed into an urban area, a small metropolitan district including the villages of Whiting, Plover and Park Ridge and expanding to the towns of Linwood, Hull, Stockton and Plover.

Stevens Point can be classified as a multifunctional city encompassing these areas: professional (especially insurance, banks, real estate), educational (university, technical school), retail trade, industrial (furniture, paper mills, printing, brewery, food processing), transportation (a large freight yard and trucking companies) and medical (hospital, clinics, medical practices). As the county seat the city had a governmental function, and because of its location in the center of the state, the city is considered a convention hub. In its central location, Stevens Point offers most of the services that are necessary in Portage County.

The U.S. Government publishes the Standard Industrial Classification Manual (SIC) which names thousands of industries and their classifications, including: mining, agriculture, trade, transportation, finances and all kinds of services and administration. Another official publication is the Standard Occupational Classification Manual (SOC) listing every lucrative occupation found in the country from graduate assistant to deckhand on a fishing vessel, from farm laborer using machinery to an electronic-sensing equipment assembler and from legislator to professional athlete, lobbyist and astrologer. By looking at classifications in both books a list of services available in the Stevens Point metropolitan area can be obtained.

Although industry is limited to a few sectors, Stevens Point has people engaged in every lucrative occupation that is necessary to the development and economy of the region.

As defined by the State of Wisconsin, Stevens Point is a third class city (10,000 - 39,000 inhabitants). The 1970 census counts a 23,479 population and the 1980 census, a 22,970 population. These figures are not reliable as there is no official register of the population. Also, some of the 9,000 university students are included in their hometown census, and others are counted as residents of Stevens Point, although they may live in Stevens Point only during the school term. The metropolitan district of Stevens Point amounts to more than 30,000 people.

Plover, located on the portage between the lower and the upper Wisconsin River, was the first village and, in 1844, the first seat of government of the present Portage County. The village was platted in 1845, but it was not until 1857 that it was incorporated as a village. When Stevens Point replaced it as county seat, Plover was disincorporated. In 1912 Plover was again incorporated but, because of financial problems, it was dissolved shortly afterwards. Since the mid-1960's a movement to incorporate Plover was again activated, and it became an officially recognized incorporated village, created out of the township of Plover (the most populated village in the county).

The western part of the county was the best access for immigrants coming from the East. There was land suitable for farming and some families settled there. In 1851 the town of Amherst, comprised of two townships, was organized. A village started along the road parallel to the Tomorrow River. It became the trading place for the area. It grew in importance when the Wisconsin Central Railroad came into the village in 1871, and a year later when the Green Bay Railroad passed about a mile to the north and had a spur close to the agglomeration. Amherst village was detached from the town of Amherst and incorporated in 1899.

The southern part of the county, because it was open, flat land with fine prairie soil, was the first area settled by farmers. Along the road from the south (now County Trunk J), a few houses were built, a tavern was mentioned in 1857, a store opened around 1860, followed by other stores, shops, a post office and residences. This became the village of Almond, detached from the town of Almond and incorporated in 1905.

In the early days, there were few settlers in the northeastern part of the county. In 1858, a Swiss named Gottlieb Staubli obtained a quarter section from the U.S. Land Office and built a log cabin on the bank of Flume Creek. (Staubli, also spelled Steubli, Stiple or even Stanbly. In the old German writing "n" and "u" are similar). He sold the land in 1861 to Theodor Staubli, went off to war and died in service. The land was purchased in 1867 by Jens Rasmussen, a Dane who moved to the log cabin and became the first permanent settler in what would be the village of Rosholt. He eventually built a dam on the creek to furnish power for a gristmill. In the 1880s a Norwegian, John Gilbert Rosholt from Waupaca County, bought timber rights and apparently made arrangements with

Rasmussen to share water rights on the pond. In 1902 Rosholt sold the timber rights to a large lumber company in Schofield and Chicago. Arrangements were made with the Chicago & Northwestern Railroad to run a branch from Elderon to transport the logs. The first train arrived in 1903. Under the name of Rosholt, the place immediately became a boom town with hotels, stores and shops, a post office, creamery, bank, warehouses and even a weekly newspaper. The newspaper was published in the latter part of 1905, but discontinued in 1906. The village of Rosholt was incorporated in 1907.

Railroad crossings were the origin of two villages, Junction City and Amherst Junction. In 1871 the Green Bay and Lake Pepin Railroad was built over the Wisconsin Central Railroad line (later Minneapolis, St. Paul & Sault Ste Marie Railroad--known as the Soo Line). In 1874 a line of the Wisconsin Valley Railroad (later changed to Chicago, Milwaukee, St. Paul & Pacific Railroad--known as the Milwaukee Road) was laid. Where these two railroad lines intersected sidings were established for transferring freight cars from one line to another, depots were built and villages started. At Junction City, the part of the highway that ran parallel to the railroad became Main Street. At Amherst Junction, the village developed south of the tracks. The street crossing railroad lines and leading to Nelsonville and to the main highway between Stevens Point and Amherst, became the main street with stores, taverns, the bank and the post office. Both villages became incorporated, Junction City in 1911 and Amherst Junction in 1912. Junction City became detached from the town of Carson and Amherst Junction from the town of Amherst.

Another village which split off from the town of Amherst was Nelsonville, incorporated in 1913. Nelsonville owes its location to a beaver dam on the Tomorrow River where pioneer Jerome Nelson, an Englishman, saw an opportunity to build a higher dam and create power for a sawmill since there was abundant woodland in the area. In 1854 Nelson bought a forty. The following year he built a sawmill on the right side of the river, probably the first in the town of Amherst. In 1870 a grist mill was built on the opposite side of the river. There were a number of Norwegian farmers in the area. Since most of the mill workers were also Norwegian, these people soon established a village. A school had been built in 1857. Later, the Norwegian Evangelical Lutheran Church was built, a post office, a bank, a creamery and stores. Nelsonville was the last incorporated village in the rural part of the county.

Park Ridge was the next village to be incorporated. It was actually a suburb of Stevens Point located in the town of Hull. An affluent residential section of Stevens Point, with business and professional people as its residents, Park Ridge was incorporated in 1938. It has remained a distinctive residential community, however, depending on Stevens Point for most of its services. It never had a post office, a school or a church of its own.

In 1947 another suburb of Stevens Point in the town of Plover was detached and incorporated as a village. Situated between Plover and Stevens Point and crossed by the Plover River and the Wisconsin River, the area has been the site of industries from the time of early white settlers. A sawmill was built there in 1852, later a gristmill and, in 1894, a papermill. The site comprises the village of McDill that was platted in 1873, and the village of Conants Rapids. The latter was platted and divided into 84 blocks but did not develop as a residential area until after the incorporation of the village of Whiting. Whiting was named after George A. Whiting, founder of the Whiting-Plover Paper Company.

Since the village of Plover was incorporated in 1971, none have been incorporated since--although a number of villages and hamlets could qualify. A few of these are true villages, with businesses along a main street, such as Arnott, Bancroft, Kellner and Polonia. Other villages formerly had some importance, offering services to the vicinity: general stores, schools, taverns, blacksmith shops, churches, creameries and railroad depots. They are now reduced to a few residences and farms; some with one or more taverns (Custer, Ellis, Stockton, Keene) or a church (Casimir and Heffron) or a grocery store (Liberty Corners, Spiritland, Blaine, Benson Corners). Others have nothing but residences and farms (Coddington, Garfield, Peru).

Kellner represents a special case. In the early 1860s immigrants (who appear to have come from Mecklenburg-Schwerin, a region of Prussia) settled as farmers on both sides of Buena Vista Creek shortly after Wood County was detached from Portage County. Along the road built on the county line, a small trading place developed--with stores, taverns, a Lutheran Church and a Moravian Church. The area grew in importance when the Chicago and Northwestern Railroad was established in 1901, with a depot at Kellner. In recent years however, the railroad line was abandoned and the tracks removed. The village, relatively close to Wisconsin Rapids, became a suburb of that city. In the telephone directory, Kellner is considered part of Wisconsin Rapids. Its main street is 80th Street South. Even Grant Elementary School is listed as located at 8511 County Trunk WW, without any mention of being in Kellner or in Portage County. The Kellner Branch Bank of Bancroft State Bank is listed at 3821 80th Street South and Kellner Klub Tavern at 4111 80th Street South.

In most parts of Europe, Asia, Africa and even pr-Columbian America people lived in villages and cities, not only for community but for protection as well. Even farmers needed protection against nomads and enemies who may have attacked isolated farms, ransacked or destroyed them and killed the inhabitants. When danger was great, localities were protected by fences or fortifications. Ancient cities have been found to be completely enclosed by natural defenses (water or cliffs) or walls with gates that were locked at night against attack. However, when white people came to settle in what was to be Portage County, attacks from Indians of the Midwest had passed. Consequently, pioneer settlers organized their farms into one lot made up of the residence, rural buildings, fields, pastures and some wooded plots. This layout was aided by the U.S. Land Survey, the township and range system which divides land in square and rectangular lots sold by land offices. This is the dispersed rural settlement, the type common to the entire country. In forests, lumbermen established camps for loggers. For farmers, and for lumbermen and their employees, there was a need for services. Towns usually began with a general merchandise store, a tavern, a post office, often a church, a school and shops of craftsmen and tradesmen. This agglomeration became the hub, or center, of the town.

Localities in Portage County developed for a variety of reasons. For example, DuBay's Trading Post began because it was established at an easy place to cross the Wisconsin River. Taverns opened at crossroads where wagons made regular stops (Ellis, Keene). A beaver dam or rapids on a stream were favorable conditions for the establishment of a gristmill or a sawmill (Jordan, Amherst, Nelsonville, Rosholt, Whiting). A railroad stop accounted for Stockton, Custer, Bancroft and Coddington; and a railroad junction was the beginning of Amherst Junction and Junction City.

In areas where there were a number of families of the same faith, it became the practice to organize a parish and, eventually, build a church. This was the origin of St. Casimir in the town of Hull; the church of Alban; the churches of North and South New Hope; St. Patrick Church in the town of Lanark; St. Bartholomew in the town of Carson; the Moravian Church near Kellner and the Evangelical Free Church in the town of Eau

Pleine. Some of these churches no longer exist. As other services were added--stores, a post office, a blacksmith shop, a creamery--some of the smaller localities became small central places (Almond, Amherst, Rosholt and Junction City). The agglomeration of Stevens Point and Plover became the main hub of the county providing almost all services needed.

Walter Christaller, a German geographer, has studied central places and established that they follow a pattern. They can be divided according to their importance from very small central places offering only a few services to large metropolitan districts offering all services. Their location does not happen haphazardly, but follows a pattern (if physical and cultural conditions are the same). A central place of first order (metropolis) is surrounded by a number of places of second order at about the same distance from the metropolis and from one another. They form a hexagon. In turn, each of these places of second order is surrounded by another hexagon of places of third order, and so on. In Germany, and in some other countries, the central place of first order is the capital of the nation; the central places of second order are capitals of states; central places of third order are capitals of provinces, etc. This is the theory, but in America's Midwest, it is only partially true. There are exceptions, but they can be explained.

In America, the administrative function is of less importance than the size of the population. For example, the central place of Washington D.C. is not as important as New York, Chicago or Los Angeles. Springfield, Illinois; Jefferson City, Missouri and Albany, New York as central places are ranked below Chicago, St. Louis and New York City. Even in Wisconsin, Milwaukee is said to be above Madison. Consequently, Milwaukee can be considered a place of second order; Madison a place of third order; Green Bay, Wausau, Eau Claire, LaCrosse and Janesville as places of fourth order. Stevens Point is a place of fifth order. In Wisconsin, instead of regular hexagons, there are irregular polygons. Stevens Point is part of the polygon comprising Wausau, Waupaca, Wautoma, Wisconsin Rapids and Marshfield. The seat of Portage County is at the center of the polygon, comprised of central places of the sixth order: Rosholt, Amherst, Almond and Junction City. This polygon could be widened to include Mosinee and Wisconsin Rapids. Each of these central places of sixth order is surrounded by places of seventh order.

All central places in Portage County originated before the era of the automobile, they served people living in the vicinity or within walking distance. Among establishments of small centers are usually found a general store, a post office, one or more taverns (some combined with a general store), school, church, creamery, gristmill, blacksmith shop, railroad station and hotel. In 1900, for example, there were 33 post offices in Portage County; now there are only ten.

A number of places where a village developed, were central places only for a while. Jordan which had been platted in 1857, consisted of four blocks with a population of 165. It had a grocery store, blacksmith shop, sawmill, gristmill, lathmill and 40 buildings--23 of which were dwellings. When the supply of timber was exhausted, the village was abandoned and eventually disappeared completely. Buena Vista was also a village with a post office and a tavern, but it was destroyed by a tornado in 1863 and never rebuilt.

Among 50 post offices established in the county, a number were in existence only for a few years. With the arrival of the automobile, most central places of the seventh order lost their importance. A number do not offer services any longer. Many stores, schools, creameries and even churches were closed. The few hamlets that survived--such as Bancroft, Custer, Amherst Junction, Nelsonville, Polonia and Ellis--remained small villages.

Plover, which became the county seat in 1844, saw the first post office in the county in 1845. It was the first agglomeration to be platted and soon had to compete with Stevens Point. The latter was platted in 1847. Its population increased rapidly and was incorporated as a city in 1858. Plover, which had a population of 450 in 1850 and 893 in 1860, saw its population decline. The 1880 census listed only 412 inhabitants; in 1890, 319. Plover is not even mentioned in the 1900 census because it had been disincorporated.

People who came to farm in Portage County first settled in the southern plain. The village of Almond originated at the junction of three county roads. Pioneers built their homes south of a low ridge at the northeast of the intersection for protection against north winds and snow. The area also provided maximum exposure to sunlight. It was about 1850 that the first settlers came. In 1857 a tavern at that location is mentioned, and a store was opened in 1860.

Pioneers also arrived in the eastern part of the county about that time. In 1858 a dam was built on the Tomorrow River to furnish power for a gristmill. Not far away was a store, and together with the gristmill the village of Amherst began and grew to its present size. It remains a central place today. The village of Junction City developed when the Wisconsin

Valley Railroad (which became the Milwaukee Road) intersected the Wisconsin Central (now the Soo Line) tracks in 1874. It soon became a central place.

The northeastern corner of the county was the latest to develop as it was far from the Wisconsin River and from the first railroad lines. The land was covered with a forest that was too remote to attract the large lumber companies.

A paper mill was built near the Wisconsin River between Plover and Stevens Point. A number of families constructed homes nearby and, in 1947, the village of Whiting was incorporated. Some time earlier people living in part of the town of Hull, situated between Stevens Point and the Plover River, wanted to be incorporated. Their petition was accepted and the village, taking the name of Park Ridge, was incorporated in 1938.

As Plover developed, the village asked again for incorporation. This was granted in 1971. New houses have been built in Stevens Point, Whiting, Park Ridge, Plover and the surrounding area. Today Stevens Point is considered the center of a metropolitan area comprising three other incorporated villages and expanding into the towns of Hull, Plover, Linwood and Stockton.

Among the other villages of Portage County, only six have been incorporated: Amherst in 1899, Almond in 1905, Rosholt in 1907, Junction City and Amherst Junction in 1911 and Nelsonville in 1913.

A few villages that never incorporated, although they could have qualified for such an action, are Arnott, Bancroft, Custer, Ellis, Kellner and Polonia. Other places which at one time had some importance and promised development, have now declined and are limited to one or two buildings. For example, Casimir, which has a church and a parish house, and Keene which at this writing has a sign reading "Keene Tavern Population - 4." Others are Alban, Blaine, Coddington, Garfield, Heffron, Peru, Stockton and Torun.

Until 1900 there was a clear division between rural and non-rural areas. Rural areas were farms, woods and swamps. Non-rural areas were cities, villages and hamlets.

With the introduction of the automobile, the situation changed. Farmers could drive to more distant places and to the city. They no longer had to go to the nearest central place to sell their produce, go shopping, go to the post office, take their children to school, attend church and visit the taverns. Non-farmers working in the city or in villages no longer had to live near their place of work, but could build a home in the country.

Rural schools closed as the city and some villages organized bus services to pick up students in the country. Some craftsmen and businessmen located their shops in rural areas, such as restaurants, taverns, beauty shops and even a branch bank.

People living in cities, not only in Stevens Point but in other Wisconsin cities and even out-of-state cities, have elected to build second homes near lakes or in the woods where they spend weekends and holidays or, perhaps, retire.

Some developers have bought tracts of country land and divided them into building lots for homes. Others have established trailer parks or mobile home courts in rural areas.

There is no longer a clear division between urban and rural areas, but a third type has been added--the rural non-farm.

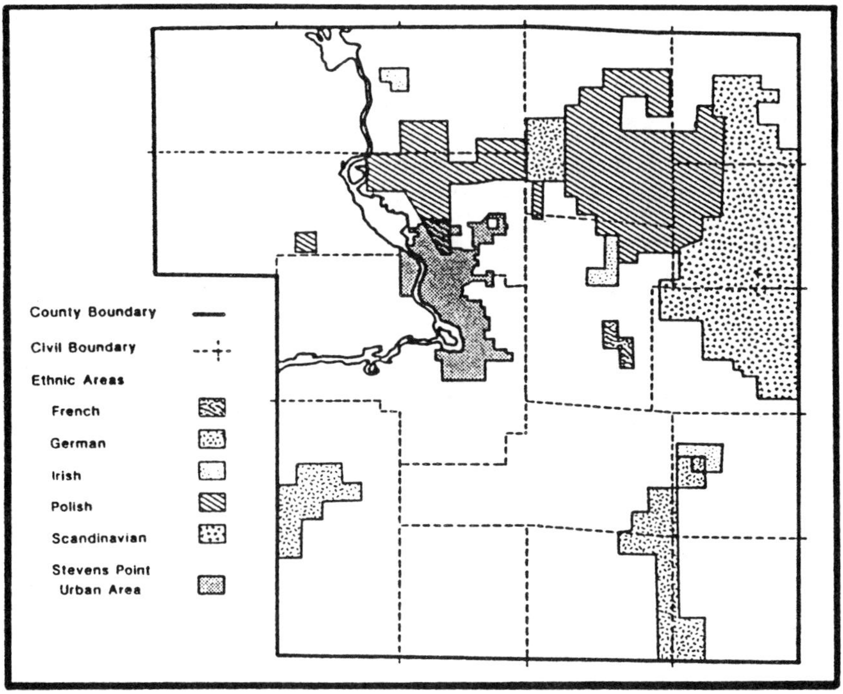

Foreign Colonies in Portage County 1876
Source: Cultural Diversity in Central Wisconsin

remain in those blocks. They are built of imported stone. Local stone from quarries, and especially from glacial erratics, continues to be used for home foundations.

To maintain high standard at low cost, brick was introduced and used for the entire construction--not only of public buildings, churches, schools, business buildings--but also of residences.

According to the 1856 census there was a brickyard in the county, but no details are known about it. In the same census, ten bricklayers and stonemasons were listed but no brick construction can be traced to that time. The most important brickyard in the area was started by a German immigrant, Edward Langenberg, who came to the United States in 1848. He settled in 1861 on a farm in the town of Hull, just north of Stevens Point, and opened a clay pit on his land to start a small-scale production of brick. But it was only in 1882 that his son, William, took over the business and started a large-scale operation. In 1884 (or 1887) the first house of local bricks was built on Main Street for Mr. Kuhl, a German who owned and operated a brewery located a block north of his home. This house, often referred to as "the castle," is very large with two towers. It is still standing today.

Actually brick building started before the operation of local brickyards and kilns when bricks were brought in by railroad. Dates of the first brick buildings are not known. The earliest known date for a brick building is 1879. The building, situated on Main Street, was a wooden frame house veneered with brick. Jules Iverson, a watchmaker from Denmark, commissioned the building to house his jewelry business. The popularity of brick as a building material began around 1880 and lasted until about 1910.

In addition to the Langenberg brickyard, other companies started by Germans or Poles, produced bricks. Most of the buildings in the central business district were built of bricks, in some instances replacing wooden constructions destroyed by fires. A devastating fire on December 28, 1874, destroyed nineteen businesses in the area between the Square, College Avenue and Third Street. In March 1885 a fire destroyed nine buildings to the south of Main Street between Water Street and the Square, involving a tailor's shop, a shoemaker's shop and six saloons.

The next three years saw an increased development in the area of the Square. A Polish architect from Milwaukee gave the Square its distinctive flavor when he constructed two prominent buildings with towers and

ARCHITECTURE AND HOUSING

Pioneers use materials that are available, easy to work with and fulfill the requirements of the land and the climate for their dwellings. Since wood was available in large quantities in Portage County, it was the first material used in building shanties for loggers and houses for settlers. Although immigrants from Europe were accustomed to stone or brick as the common building material, Americans from the East (Yankees) and Canadians were already used to frame houses. Wood thus became the predominant type of construction material. A number of immigrants from Central Europe and the Scandinavian countries who settled in wooded areas, built log cabins. In Ellis, in the middle of a prairie where there was an outcrop of sandstone, quarries were opened. Sandstone was extracted and used for some buildings. At this time, there exists two buildings of sandstone, one in the center of Ellis is no longer inhabited and is used for storing hay. The other, northwest of Ellis, is the residence of a farm and is still in good condition.

In addition to the Ellis quarries, other quarries were opened in the county. In the north quarries furnished granite and quartzite, both hard rock difficult to cut. In the south, especially on Mosquito Mound, were quarries that furnished sandstone. At one time the quarries were used extensively, but are now abandoned.

In areas covered by moraines, there are numerous stones of all sizes. They were too hard to be used for building entire houses, but because of their strength and insulating properties, these stones were used for chimneys, foundations for houses and on farms for stabilizing parts of barns and milk houses. In most cases the use was limited to walls formed of glacial erratics that needed little cutting and trimming and could be assembled even without any mortar.

Stone, because it was stronger, more durable and more fire resistant than wood, was also used for public buildings. Though more expensive, requiring specialized workers for quarrying, cutting, transporting and building, stone was considered of a higher standard than wood. Consequently, stone was used for some businesses, especially banks and public buildings. The earliest stone building in the county was the jail and county court house erected in 1868.

Three blocks of stone business buildings are grouped on Main Street near the Square. One building is dated 1871. Only four old stone houses

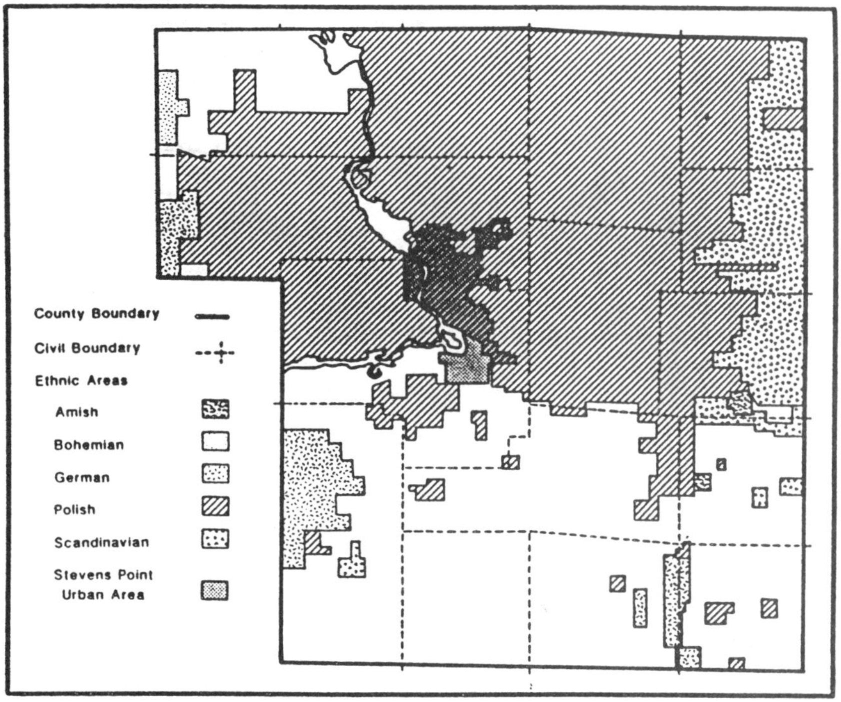

Areas of Non Anglo-Saxon Ethnic Groups in Portage County, 1972

Source: Cultural Diversity in Central Wisconsin

cupolas. In 1893 the Opera House was built among businesses and banks on Main Street.

It was during the same period that a number of public buildings were erected in brick: St Joseph Church and School in 1883, Jackson School in 1887, Garfield School in 1892, Normal School in 1894, St. Peter's Church in 1897 and the German Evangelical Friedens Church (United Church of Christ) in 1887. The new Sacred Heart Church at Polonia, which replaced the smaller church, was dedicated in 1902. At that time Sacred Heart was considered the largest rural church in the county. This church however, was struck by lightning which set the building on fire and gutted it on St. Patrick's Day, 1934.

Brick was also the material used for most warehouses, factories and buildings connected with wholesale products and railroads.

Although brick was used for walls, stone was often used in conjunction with brick--especially for jambs, sills, lintels of doors and windows and for foundations at or below ground level in most brick buildings. For contrast, stone was combined with brick, as at Nelson Hall and the former Campus Laboratory School, on the University of Wisconsin-Stevens Point campus.

Outside Stevens Point, brick was used in business construction on the main streets of villages, incorporated or not, and in central business districts of Almond, Amherst, Amherst Junction, Arnott, Bancroft, Kellner, Junction City, Nelsonville, Plover, Polonia and Rosholt. In Stevens Point, as in big cities, buildings ran from one end of the block to the other, sometimes with only a few inches in between. In villages only two, three or four houses were in a block. Brick houses in those commercial blocks can be traced to a period from 1880 to 1910.

In Stevens Point, brick was also used for a number of residences. They can be grouped in three zones: the band stretching eastward from the central business district to the university along Main Street and parallel streets, a small cluster centered around the courthouse and a large zone surrounding the southside business district and the repair yards of the Wisconsin Central Railroad. The first section represents the natural eastward expansion of the city where not only homes, but churches, schools and the hospital were built from brick. The cluster around the courthouse is probably an expression of affluence since many homes are quite large with ornate lintels. A number of them probably replaced pre-existing wooden structures. Another zone with a number of brick buildings is the northwest residential district with St. Peter Church as center. Surrounding are the nuns' house, the rectory and the parish school. Nearby is Kosciusko Hall, built in 1909, which has a co-op store on the ground floor and on the second floor a large hall for meetings. Not far away is St. Joseph Convent. This part of the city was originally considered the Polish-speaking community.

The period when brick was used as the major building material lasted about 40 years, approximately two decades before and two decades after the turn of the century. The Langenberg brickyard was listed in the 1923 directory but is not found listed any more after that.

Brick was used primarily in the northwest district where Polish people were concentrated and on the southside where Germans lived in large numbers, attracted there by work in the railroad shops. In the central zone from the courthouse to the university, the Polish and the Germans were joined by Yankees and people of other ethnic groups in selecting brick as a building material.

Wood, however, has always been the major material used in building. The first dwellings were shanties or plain houses without any style, but soon wealthy people built stylish homes reflecting the fashion of the time.

Since colonial times home styles can be identified according to specific characteristics. At the same time some builders were faithful to older styles or modified the current trend.

By 1880 five basic American domestic architectural styles could be identified: Colonial, Georgian, Federal, Greek Revival and Victorian. Often there are differences from region to region, traceable to the climate or to the origin of the settlers.

The Colonial style, from 1690 to 1760, is a two-story house with symmetrical facade, an equal number of windows on each side of the main entrance and a single chimney in the middle of the roof. There are varieties such as the Cape Cod, one story high, and the Salt Box with two stories on the front and on the back and a roof starting above the first floor.

The Georgian, from 1720 to the Revolutionary War, is a two-story building with two chimneys. Identifiable are elements of Classical Roman architecture such as cornices, columns and pilasters. In time these features became more numerous and more elaborate.

The Federal style, 1780 to 1820, follows and is an outgrowth of the late Georgian style, but more austere. It includes a porch in front of the main entrance. This style was not popular, and from 1815 to 1840, the Greek Revival style became dominant. It exemplifies a Greek temple with columns on the front and a gable on the facade. Generally the style is simplified to a colonial house with some Greek elements as pilasters or a massive porch around the front door.

Although these styles developed before Portage County was settled, they were often used by architects for houses in Stevens Point and in villages, as well as farm houses.

Greek Revival was followed by the Victorian style, a term that applied to the many styles that appeared during the long reign of Queen Victoria (1837-1901): Gothic Revival, Italianate, Second Empire, Stick style and the Queen Anne style. The Victorian movement was a trend away from the classicism of Greek Revival and Colonial architecture.

The spread of these styles was due largely to the influence of A.J. Downing who published a first book "Treatise on the Theory and Practice of Landscape Gardening." Included was a long chapter on "Landscape of Rural Architecture" which went through more than 15 reprintings. It was followed by "Cottage Residences" which had 12 reprintings. Finally "The Architecture of Country Houses" was published in 1850, two years before the death of the author. It had nine editions. A Swedish lady who visited the county in 1849-1851 wrote that "nobody, whether rich or poor, builds a house without consulting Downing's works. Every young couple who sets up housekeeping, buys them." Many books and periodicals devoted to architecture have been published since then.

Downing divided country houses into three classes: cottages, farmhouses and villas. He considered a cottage to be a dwelling of small size intended for the occupation of a family who manage the house themselves or at the most, with the assistance of one or two servants. "The majority of such houses are occupied by working men who own the ground upon which they stand, build them for their own use, arrange them to satisfy their own peculiar wants and gratify their own tastes. The cottage is to imitate, as closely as cheap and flimsy materials and a few hundred dollars will permit, the style and elaborate ornament of the villa with its expenditure of thousands." By this Downing considers all houses built in suburbs or in small towns for people working in businesses, offices and shops, as opposed to either farmers or to wealthy people. Farmers live in farmhouses that are appropriate for their needs and can be beautiful without imitating the style or decorations of the ornamental cottage or villa. The villa is a country house of large accommodations, requiring the care of at least three or more servants. With one or two exceptions, all houses in Portage County built before 1900, respond to the definition of a cottage.

Downing is regarded as one of those responsible for discrediting the Greek Revival style. For him it was extravagant and inconvenient. As he had no formal training in architecture, Downing's models of houses are largely inspired by the early types of Victorian style; the Gothic and the Italianate with the principle of irregularity or more "natural" forms. He felt a building should be viewed from many angles and each, to a degree, should be unique.

The Gothic style is patterned after medieval Gothic structure, though largely modified. Emphasis is on the vertical effect, windows are tall and end in pointed or rounded arches. The most conspicuous feature is often the bargeboards ornately carved which gave rise to the term "gingerbread" houses. Many have verandas.

The Italianate style is inspired by Italian country villas. Windows are tall and often round-topped. Often there are balconies, porches, bay windows and towers or cupolas. The roofs are low-pitched or flat.

Second Empire, referring to the French Empire of Napoleon III who reigned from 1852 to 1870, was the most prominent style used between 1860 and 1880. It is characterized by a very steep roof sloping to the wall with a flatter portion above it. In the roof are windows which allow the attic to be used as a separate story. These windows are actually "mansardes," named after the French architect who started to use them at the time of Louis XIV. These houses have various ornaments: porches, towers, tall chimneys and ironworks.

The Stick style is a partial derivation from the Gothic form and the Swiss chalet. Exposed stick work is the most conspicuous feature. This amounts to uncovering framework, or projecting beams and studs to the outer wall. The roof is steep. There may be verandas in one or two stories.

Queen Anne style is a misnomer. It is somewhat a reflection of Jacobean and Elizabethean architecture. It exhibits extremely complex designs in terms of shape, texture and color. Wall openings come in many shapes and sizes. Gables, turrets, oriel windows and porches are used freely.

The Queen Anne style was a transition from Victorian to Eclectic concepts. Eclectism that started around the turn of the century is more an age of selective borrowing than a style in itself. It looks at architecture of the past and transforms suitable elements into new designs. It was also a period of adapting architecture to the rapid changes in construction, in materials, engineering and building techniques. Among Eclectic types, some are inspired by American styles such as Georgian Revival or the Neo-colonial, but others were imported from Europe such as the English Tudor, Jacobean or Jacobethan Revival (after the style under King James I and Queen Elizabeth), French Provincial, Neo-Dutch Colonial and Neo-classical. For example details that evolved from a "gingerbread" mode through an Egyptian design might be conspicuously expressed in pyramid-shaped porch supports. In some cases, a sun porch is created by completely enclosing the area in glass.

Styles are also imported from the Southwest; the Mission style, inspired by the old Spanish missions in the Southwest and in California and the Pueblo style, mixing Spanish and Indian elements and represented in the Southwest by houses of massive appearance, minimum ornamentation and flat roofs.

In addition to the Eclectic style, another style developed in America during this century is the Prairie style. A conception of Frank Lloyd Wright, the Prairie style is in harmony with the broad, gently rolling terrain of the Midwest. The emphasis is on the horizontal. Roofs are massive and low pitched, and windows may form long series.

Another type is the bungalow. Originally inspired by houses built in India by the English, the bungalow is a one-story building which stresses the horizontal in simple lines.

Since World War II, five styles can be found to have been built: the ranch, in some ways similar to the bungalow; the Cape Cod; the Neo-Colonial, generally a two-story house; the split-level house which appeared in the early 1950s and the flat-roofed functional modern house. These houses were adapted to the automobile era with a garage for one or two cars incorporated into the design. However, sometimes included was a covered area on the side of the house under which the car was stored.

Houses of almost all styles have been built in Stevens Point or in other parts of Portage County. A number of older ones have been torn down to be replaced by modern construction or to make way for traffic or new developments such as highway bypasses or malls. Many have been altered by additions, elevation, destruction of some aspects, opening or closing doors and windows.

Until the middle of this century, all houses, except those in the business districts, were built as family homes on one lot. The majority were frame houses but, as brick and stone were more prestigious, occasionally walls had a veneer of stone or brick. Sometimes a simple tar paper imitating brick was used. They were all of one, one-and-a-half or two stories, often with a basement that could be used for many purposes: storeroom, washroom, workshop or recreation room.

In recent years a number of apartment houses have been built; some within residential areas on Main Street and vicinity and others on the outskirts of town. Some apartments are large two-story buildings with several units on each floor and a front and back entrance. Others are duplexes, while some are rows of single apartments one or two-stories high and having a separate entrance. In the suburban areas and in the villages, the architecture has followed the general trend of American houses. Here and there, however, houses can be found that have not been modified since they were built more than 100 years ago.

Examples of most American styles of architecture, fashionable since 1850, can be found in Portage County. Although the majority of houses are plain dwellings with four walls, a roof, windows, doors and perhaps a porch as protection against rain or wind, often details can be found typical of a particular style or period. Such characteristics may be evident in ornaments or in the disposition of windows, doors, roofs or chimneys. Some houses however, show a definite desire of the builder to follow the fashion of the day, or to express his taste or the taste of the person who commissioned the building. A number of houses show elements of two or more different styles in the same structure.

There is a house in Plover that is the work of the famous architect, Frank Lloyd Wright. The property is fenced and surrounded by hedges and trees so that those on the outside can only have a glimpse.

It is believed that the oldest house in Stevens Point is at 1924 Main Street. A one-story square building, it is said to have been erected in 1855. There is no proof of that date, however. Another house claimed to be the oldest is at 1516-1524 College Avenue. It first stood at the northeast corner of Main Street and Smith Street, but in 1908 the house was moved to its present site. It is two stories high, of Federal style but with features of Greek Revival (pediments) and Neo-Classical (front porch).

Few houses have the characteristics of one style. A number of houses may have followed the rules of one style when built however, when they were enlarged or remodeled, elements of other styles were added.

Houses representative of one style can be found in the following list:

Greek Revival: The house that is now on the grounds of the Portage County Historical Society in Plover.

Victorian Gothic: 1924 Pine Street

Victorian Italianate: 1902 and 1918 Pine Street, 2027 Lincoln Avenue, 1354 Strongs Avenue

Victorian Second Empire: 1416 Main Street

Victorian Stick: 2101 Church Street

Eclectic Queen Anne: 1100 Brawley Street

Eclectic Georgian Revival : 1400 and 1909 Clark Street

Eclectic Jacobethan: 1202 Illinois Street

Eclectic French Provincial: 1501 Clark Street

Eclectic Cube Egyptian: 424 Prentice Street

Eclectic Neo-Dutch Colonial: 1547 Pine Street

Neo-Colonial Cape Cod: 2317 Jefferson Street

Neo-Colonial Salt Box: 2048 Dixon Street

Prairie: 1417 Main Street

Mission: 1515 Frontenac Avenue

Pueblo: 2048 Dixon Streets

These are but a few examples. There are many other houses representative of one style, not only in Stevens Point but in the villages and in rural areas.

Wooden frame buildings have continued to be the main type of construction in Stevens Point, in the villages and in rural areas. Brick and stone were used for foundations, chimneys and as decoration. However, a few public and business buildings have been built of stone since the 1880s. The Episcopal Church of the Intercession was the first stone church built in Stevens Point, 1892. Sandstone from a quarry near Eau Claire was used. Only two other all-stone buildings were built before the 1930s: St. Stephen Catholic Church built in 1920, of granite foundation, sandstone jambs and walling, and Sentry Insurance office building constructed of limestone. The Sentry building was built in 1922, with an addition in 1930. For the most part, stone was used in conjunction with brick, for example: Nelson Hall, the first dormitory of the university, was built in 1915 and Campus Laboratory School, 1929. In the 1930s, buildings built in stone were: St. Paul Lutheran Church, 1935, and P.J. Jacobs School, 1938.

Brick was the main material used for the walls of other buildings: Normal School, 1894, and its additions, the west wing in 1901 and the east wing in 1915. Most other schools incorporated brick walls. Other buildings with brick material are: St. Michael's Hospital, 1913; Whiting Hotel; St. Joseph Catholic Church; Presbyterian Church (now Frame Memorial United Presbyterian Church); St. Stanislaus Catholic Church; Bible Baptist Church; First Baptist Church and Peace Evangelical Church of Christ. In the villages most of the Catholic Churches are also brick buildings: St. Bartholomew, Mill Creek; St. Casimir, the Town of Hull; St. Bronislava, Plover; St.John, Heffron; Sacred Heart Church, Polonia; St. Adalbert, Rosholt; St. Mary, Torun; Immaculate Conception Church, Custer and St. Mary of Mount Carmel Church, Fancher.

Since World War II, new techniques have changed the traditional way of building. Different materials were used that had been of little importance in construction in the area. Materials used included cement, concrete, iron for framing and siding and glass. At first these materials were used only for large public or business buildings. Later the material was adapted to use in the building of private homes.

Modern styles of architecture are especially impressive as exemplified in the headquarters of Sentry Insurance Company and several university buildings. These university buildings include the Fine Arts Building, the Albertson Classroom Center and the Learning Resources Center. Other modern buildings are Trinity Lutheran Church, St. Paul Methodist Church, Faith Lutheran Church in Junction City, St. James Catholic Church in Amherst, St. John Lutheran Church and United Methodist Church in Almond. Among other recent buildings worth mention are the City Hall in Stevens Point, the Plover Hall and CenterPoint Mall.

Change in life-style has also brought a change in settlements. Until the middle of this century, almost all residences were family homes. In the county, two new forms began to develop: apartment buildings or complexes and the trailer or mobile home courts, which were already common in large cities and suburbs.

The first modern apartment complexes opened in 1964--Feltz Apartments on Fourth Avenue and Oxford Apartments on Vincent Court. Many others followed. Apartments rapidly became popular for several reasons. Many older people often find it a burden to keep a house and to maintain the accompanying yard. Also, students who are here for only a few years find advantages to living with companions in apartments--it gives them more freedom, allows them to prepare their own meals and does not require as much upkeep as a house. Newly-married couples also find apartments convenient since they usually intend to save money in order to buy a home of their own. Other candidates for apartment living are people who have temporary work contracts (university faculty, employees on limited assignments or large construction projects). They usually have no intention of settling permanently in the area.

Some apartment houses consist of only one building but most are part of a complex of several buildings, either attached or completely detached. Most are one or two-stories high but the Hi Rise Manor has seven stories which makes it the highest apartment in the downtown area. In some complexes each apartment forms a single or two-story house with an outside entrance; others have all apartments opening onto a central corridor. Some complexes have only a few units but others, namely The Village, have more than 100 units. One special type of complex is the Honey Comb Apartments consisting of five hexagonal buildings divided into six apartments, each with a triangular shape. Inside, each has a sitting room, bedroom, kitchen, a partial sitting room that can be separated by a screen and a mezzanine. Of similar design are the National Apartments in the same area.

Most complexes have a two-bedroom apartment, or both one and two bedrooms--but the Honey Comb Apartments have only one-bedroom apartments. The Madison View Apartments have one, two, three and four-bedroom apartments. Almost all apartment complexes have stoves and refrigerators and, thus, are considered partly furnished. Only a few, The Village and the Honey Comb Apartments, are fully furnished. Most complexes offer laundry facilities. Among other amenities offered by some are garages, storage spaces in basements or attics and gardens. The Village provides a swimming pool, ping-pong tables and basketball courts. Some use electricity while others use gas for heat. Utilities are sometimes included in the rent but in other complexes they are charged directly to the renters. Many complexes are owned by companies, some located in Milwaukee; others are owned by individuals. Apartment complexes are generally considered to be good investments.

In a special category are Hi Rise Manor and Edgewater Manor. They are owned by the Stevens Point Housing Authority and are intended for low-income, elderly people who are on Social Security. Both are near downtown Stevens Point. Within walking distance (since many of the tenants have no car) are CenterPoint Mall, shops of all kinds, a number of churches and the Lincoln Center, a meeting place for elderly people.

In most cases, location determines the types of residents. Those units close to the university have many students and some faculty. The Village caters mostly to students since it is located near the university and offers what students want. Rates are by semester and special rates are available in the summer. Some complexes do not rent to students and some complexes, in addition to Hi Rise Manor and Edgewater Manor, have primarily elderly people as tenants.

Many complexes are built on the city outskirts since they required large spaces and, when they were planned, no such spaces were available close to the downtown area.

Trailer parks and mobile home courts began as spaces made available to people traveling with a trailer to spend a night or two. But eventually they became residences for people who were transferring to a new place of work, or who had a temporary occupation away from home. In the latter instance, trailer courts were set up to provide utilities such as electricity, water and sewer or septic systems. Some court owners purchased trailers and rented them. Later, in addition to trailers, mobile homes were accepted into the courts. Mobile homes are similar to trailers but without wheels. They are simply prefabricated houses. Consequently most courts are called mobile home courts and some have taken the name of village. Trailer parks and mobile home courts are subject to strict regulations. Before they can be established, the site has to be acceptable to the Wisconsin Department of Natural Resources and the Department of Public Health. These regulatory bodies look especially at the drainage of the area, soil composition, possibilities of pollution and other ecological factors. If accepted, the State may approve granting a license which is given by the city, village or township in which it is located. The state assesses a tax.

Newby's, in Plover, was the first trailer park established. It began in 1953 with three trailers. For years it was the only one, but in 1960 Fairview Village started with 12 trailers in the Town of Hull, close to Highway 66. Other mobile home parks were started during the 1960's. At this writing there are 23 licensed mobile home parks in Portage County. Although the number of parks is almost the same as 15 years ago, the number of units per park has greatly increased. In 1971 there were 550 units; in 1984, there were 1,221. One court, Horner's, which started in 1967 with 11 trailers, is no longer considered a mobile home court but has taken the name of apartments. The size of mobile courts, according to the licenses, ranges from a single unit (Ridgewood in the Town of Plover) to the largest court at Plover Pine Village with 235 units (increased from 60 in 1971). The one unit at Ridgewood is explainable because camping with tents and recreational vehicles is the main object of the place and it is, in fact, called Ridgewood Campgrounds. Second largest of mobile courts is Recre Acres with 211 units (up from 56 units in 1971) and Evergreen Villa with 101 units (against 38 in 1971).

The majority of mobile home parks are located in the metropolitan area of Stevens Point--three in the city, one in the Village of Whiting, five in the Village of Plover, four in the Town of Plover, three in the Town of Hull and two in the Town of Stockton. Only five are located outside: one in the Town of Dewey (Crocker's Landing) one in the Town of Junction City (Sunset Trailer Court) one in the Town of Amherst Junction (Hedbany's) and two in the Village of Bancroft (Town of Pine Grove). All are located close to a major highway leading to Stevens Point.

Trailer parks and mobile home courts differ in organization, amenities and occupants. Some were built in open fields; others in wooded areas or surrounded by trees and shrubbery.

Mobile homes are usually laid out in three patterns: a single main driveway with homes on both sides, or homes on one side only; the gridiron pattern in which the trailers are within blocks and the horseshoe pattern where trailers form one block and a driveway at the entrance gate continues around the block and ends at the exit gate. Several courts combine patterns.

Some courts are well organized and have strict rules. Mail boxes may be grouped together forming a line near the highway. Rules may include: no pets, no plantation and an obligation to keep the grounds neat. Others allow plantation and landscaping, even the establishing of small gardens, and pets. Several camps offer amenities such as areas with play equipment for children, sports areas and swimming pools. Some are owned by companies or owners not in residence; others have the owner staying on the property. Some rent out a few or all the mobile homes; others rent only the space for a trailer or a mobile home. These differences explain the variety of residents in the courts.

Reasons for living in a mobile court are about the same as those for living in an apartment: people who expect to be in the county temporarily (students, workers), retired people or families who moved to the area but could not afford to buy a house. Many of the latter get accustomed to living in a mobile home and remain, discovering that life in a mobile home court can be as pleasant and convenient as living in a house in town. Consequently, some mobile home courts have taken on the atmosphere of small towns where everyone knows everyone and where residents have good relations among themselves.

In recent years a new form of settlement has appeared: the subdivision. Development of subdivisions seems to be related to families having two or more cars. This results in people not being limited to certain locations but can drive freely to work, school and stores. This migration to the suburbs was evident some time ago in the large cities, but in Stevens Point it began after World War II. It grew especially in the '60s and '70s. Developers or companies buy land and divide it into lots that are sold to people who plan to build. Subdivisions are not far from a city and are close to a main highway. Some are along the Wisconsin River, some are in open country and others are in wooded areas. Some have roads that access every lot. In some wooded areas, lots are large and surrounded by trees. This provides home owners with the impression of living in the wilderness with complete privacy.

RELIGION

Even from the early days of settlement in the Midwest, religion played a role in the lives of pioneers. Many had been members of a church in their former town, and as they came to the Midwest they formed a congregation of the same religion.

The year 1847 saw the first recorded time religious services were held in Plover and Stevens Point. A Methodist preacher conducted the services. In 1849 a Presbyterian missionary on his way to Wausau, lamented in his diary at the revelry and profanity resounding through the air in the then frontier town of Stevens Point. His views were shared by a pastor who had come from Norway, a man who conducted the first Lutheran services in Stevens Point and lodged at the Raftsman Tavern. In a letter to his family in Norway he wrote of the logging crews who stayed overnight in Stevens Point: "...the scum of humanity, the dregs of both Europe and America who live a life of constant drinking, gambling, swearing, and even of occasional murder."

Most residents of Stevens Point however, were not of this mold. They gathered in private homes, according to their faith, and held religious services. When a schoolhouse was built, religious meetings were held there.

In October 1852 General Ellis said the morning prayer for the Episcopal Church in his home in Stevens Point. In December of the same year a congregation was formed, the first one in the village. Ellis was a prominent man. He had been Surveyor General, from which he kept the title of General, and he was head of the United States Land Office in Stevens Point. The Episcopal parish counted seven communicant members. Two years later a church was built, the first in the county.

In 1851 Father Godhart, a Franciscan priest from Fort Winnebago (now the city of Portage), visited Stevens Point. At the time there were only three Catholics. But with the arrival of some Irishmen, a Catholic group was formed, and in 1853 the first mass was celebrated in the schoolhouse. Father Godfrey from St. Mary in Green Lake County came to attend to the religious needs of the group. Masses were celebrated in private homes until St. Stephen's Church was dedicated in 1856. The preceding year, Congregationalists had built their church.

The First Baptist Church was organized in 1855, and regular meetings were begun in 1857 when a resident pastor arrived. The first baptism by immersion in the Wisconsin River was held in 1858, and the following year a building was dedicated. The Methodist Episcopal Church was built in 1857, and the Presbyterian Church in 1858. Both Methodists and Presbyterians also built churches in Plover.

All church buildings were of wood, some very small and primitive. When the congregation could afford it, churches were replaced with better buildings, mostly of brick and stone.

The First Methodist Episcopal Church was incorporated in 1866. Later they acquired the Presbyterian Church in Plover. The First Presbyterian Society of Plover and Stevens Point was incorporated in 1856 and, after Stevens Point became the county seat instead of Plover, it was considered that the church in Stevens Point was sufficient for the area. The building in Plover, which in 1887 became the Plover Methodist Episcopal Church, was used until 1965 and is still standing. It is the oldest public building in the county. It is now part of the museum of the Portage County Historical Society.

All church services were in English, except in the Catholic Church where mass was in Latin, but confessions and preaching were in English.

Churches in Stevens Point and Plover were attended by people living in the vicinity, but many families lived too far away to attend services regularly. Some new settlers did not speak English and some newcomers still maintained a connection with a religious denomination in Europe. In various places in the county, groups met in private homes or in schoolhouses to practice religion in their native language.

In the town of Sharon, Germans from the Palatinate and the Rhine region, as early as the 1850's, held meetings related to their Catholic faith. In 1857 St. Martin Church was built at a site that would later be Ellis. English Methodists held meetings in the towns of Almond, Buena Vista, Belmont and Amherst.

Scandinavia in Waupaca County, having developed as the center of the Norwegian colony, formed a Lutheran parish affiliated with the Norwegian Synod, a synod organized in 1853 in southern Wisconsin. Early Scandinavian settlers in Portage County attended services in Scandinavia.

German Methodists formed a congregation in the town of Almond. Irish settlers formed Catholic congregations in the towns of Stockton and Lanark.

Polish immigration began in 1858. The first group settled in Sharon and attended St. Martin Catholic Church. They did not understand neither English nor German. Fortunately, a Polish priest became the rector of St. Stephen Church in Stevens Point and he visited fellow countrymen. In 1864 the Polish were permitted to form their own parish, and the following year they built St. Joseph Church. The church was erected near an intersection known as "Polish Corners" where there were four saloons. The owners refused to close the saloons during mass, and church services were often disturbed by unruly and noisy customers. A new pastor, Father Dombrowski recently from Europe, was appointed to St. Joseph's. As the agitation over the saloon disturbances continued, Father Dombrowski proposed that the church be moved to a more favorable site. With the approval of the Bishop, the church was moved in 1872 two miles east to a hilltop Father Dombrowski called Polonia. A few parishioners stayed behind and built a large church in Ellis, St. Maria. Excommunicated priests serving as pastors caused the church to be closed after a few years. It became known as "the condemned church" and remained empty until 1918 when it was demolished. The church in Polonia was first under the patronage of the Annunciation of the Blessed Virgin Mary, but in May 1875, the church was destroyed by a fire probably set by incendiaries, the elements of "Polish Corners." In September of that same year, the cornerstone of a new church was laid under the name of Most Sacred Heart of Jesus, commonly called Sacred Heart.

Between 1860 and 1875 the county population doubled, and from 1875 to 1895 it doubled again to 28,531. During that time many congregations emerged and several of those already in existence built new churches, enlarged standing churches or rebuilt primitive ones. New parishes were created.

In Stevens Point Presbyterians were divided between the old and the new school. In 1864 a conference with the Congregationalists and the Presbyterians resulted in the merger of the three groups and the organization of the First Presbyterian Church in Stevens Point. Eventually they built a new church, begun in 1892 and completed in 1894. It was named Walter R. Frame Memorial Church in memory of the Reverend Frame who had been pastor from 1884 to 1890.

St. Stephen's parish began modestly. Members increased and became important with the arrival of the Irish, who left their country because of the Great Famine, with the number of German- speaking Catholics from the district of Trier and from the districts of Moenchhof and Wieselburg in the Austro-Hungarian Empire. Later Germans came from Cologne, the Rhine Province, Bavaria and Silesia. They were not farmers but laborers attracted by the railroad. Others started businesses of their own. St. Stephen's became so large and mixed in languages that the pastor advised the Germans to form their own parish. They did, and created St. Joseph parish. In 1884 they built their own church closer to the railroad depot, and in the neighborhood where most German-speaking families were living.

Polish families had also settled in the city. They were not welcomed by the Irish who considered St. Stephen's as their church. At first the Poles were not admitted as members of the congregation, and when they came to Mass, they had to sit in a separate section. In 1876 they demanded to be able create their own parish. Their request was accepted and they formed St. Peter parish, and in 1876 they built the church that is in the northeastern part of the city where most Polish families have settled.

The Scandinavians were Lutheran and a few of them were living in Stevens Point. From time to time a pastor from Scandinavia came to visit them. A number of Germans were also Lutheran, and in 1872 the two groups formed a congregation. The Scandinavians met with the Germans and it was decided to build a church that would serve both groups. In 1874 Trinity congregation was officially created, and the Norwegians (it seems they were the only Scandinavians) bought half the interest in the church that the Germans had just built. For 24 years Germans held their services on Sunday mornings and Norwegians on Sunday afternoons. But in 1897, the two groups considered a separation and new buildings. In 1898, Trinity Church for the Norwegians and St. Paul Church for the Germans, were built.

All German Protestants were not Lutheran. The German Methodist Episcopal Church was founded in 1888. In 1895 12 Germans organized Evangelische Friedens Gemeinde Stevens Point (Evangelical Peace Community). At first they had services in Zimmer's Hall on the south side, but in 1897, the church in the vicinity of the railroad depot was dedicated.

In 1900, Stevens Point with a population of 9,524, had 12 religious congregations: 6 were English; 4, German; one, Polish and one, Norwegian. The services were held in the congregations' native languages, except the Catholic Masses were in Latin.

All the churches at that time were Christian, but there were Jews in Stevens Point who also wanted to have their congregation. In 1904 Beth Israel Congregation was organized with members from communities throughout Central Wisconsin. The synagogue was built in 1905. The rabbi taught Hebrew to the children and held services in that language.

Since children were educated in English, many spoke English with each other and sometimes at home since some parents had also learned the language. Generally the third generation did not even speak the tongues of the grandparents, and eventually English was substituted for other languages. In 1913 English replaced Norwegian in the Trinity congregation. In Friedenskirebche, which was known as Peace Church, services were held in German until 1920. At St. Peter's, sermons were still given in Polish.

Settlers outside of Stevens Point began their congregations using their native language, but later switched to English.

When the United States became independent, the main religious denominations were Methodist and Baptist. Presbyterians, Episcopalians and Congregationalists also had sizeable churches, and there were a number of Lutherans and Catholics. It is, therefore, not surprising that Yankee Methodists started church groups in Portage County, followed by Presbyterians and Baptists. As stated earlier, the first religious sermons in Plover and Stevens Point were preached by a Methodist minister in 1847. By the 1850's, the Methodists were active in the southern part of the county. In Almond there is mention of a Methodist conference in 1857 and a Methodist Church in 1859. In the town and village of Almond there were German settlers, attracted by a Methodist revival service, who joined the Methodist Church. Eventually two German Methodist churches were built, one north of the village and the other east.

In 1855 in Amherst, a Methodist Evangelical Group is believed to have existed. The Methodists had groups in various places, holding meetings in private homes or in schoolhouses. In the town of Belmont in 1875, Blaine Church was declared a community church, open to all Christian denominations. However, it appears that only the Methodist Episcopal congregation and, for a time, the Danish Lutherans held meetings there once a month. Outside of Blaine, two other Methodist churches were built in the town: Grant church in 1886 and, after several revival services, Dopp church in 1906.

Baptists also held religious meetings in homes and schoolhouses without forming regular congregations. In 1901, when railroads had depots at Almond and Bancroft, the American Baptists Home Society sent chapel cars where Baptist services were held for groups that had no church. Eventually Baptist churches were built in both villages.

Many Norwegians who were mainly Lutheran came to Wisconsin. By 1853 they formed the Norwegian Evangelical Lutheran Synod.

In the town of Buena Vista two Methodist churches were built in 1875, one at Keene and the other at Liberty Corners. Scandinavia soon became the largest congregation of the Norwegian
Evangelical Lutheran Church in the United States.

When Norwegians first settled in Portage County, they were members of the Scandinavian congregation. Before long they wanted their own congregations. They organized in New Hope in 1857, and built a church in 1870. The congregation became the second largest in the country. In Amherst the congregation was organized in 1877, and Peace Norwegian Lutheran Church was built in 1879. In Alban the congregation took the name of Norwegian-Danish Evangelical Lutheran Church since there were several Danish families in the area. A church was built in 1878. A Norwegian Evangelical Lutheran congregation was formed in Nelsonville and the church was built in 1888.

In 1887 a split occurred in the Norwegian Lutheran Church which affected churches not only in Portage County, but in other counties as well. The New Hope Church was divided. One group kept the old church which became North New Hope; the other group built a new church on the same road about a mile away, South New Hope. Alban Church was also affected by the split. Those who left built Concordia Church.

Many Germans were Lutheran. They started congregations and built churches such as St. Paul in Amherst, 1860; St. John in Kellner, 1872 and St. Paul Lutheran in Buena Vista, 1884. A Lutheran church was built in the county north of Almond but was moved to the village in 1918. A Danish Lutheran congregation was founded in the town of Belmont. German Lutherans had meetings in homes or chapels in the towns of Carson and Eau Pleine. Eventually they formed a congregation and in 1918, built St. Paul Lutheran Church.

Other denominations started congregations and built churches. In Almond Seventh Day Adventists built a church near Spiritland in 1890. It was later moved to Almond. A mission of the Protestant Episcopal Church in Amherst was established and St. Olaf Church was built. There were also groups of Evangelical churches dispersed throughout the county. One of them started as a Swedish Evangelical Church, north of Junction City. It still exists under the name of Evangelical Free Church.

The large numbers of congregations and religious groups were due to the fact that people were isolated with only a few good roads. Few people had horses and buggies to drive to the churches of their faith and of their language. As religion was an important part of their lives, most early settlers walked to places of worship near their homes, whether or not they were of their faith or language.

In Kellner there were Germans of the Moravian faith. The Moravian Church is one of the oldest Protestant churches dating back to the movement of Huss in Moravia. As the Moravians were persecuted, they moved to Germany where they had many followers. With time, they became a German-speaking church. Moravian missionaries came to Georgia as early as 1735. At present the Moravian Church in America counts about 60,000 members.

Some members of the First Moravian Church were in Grand Rapids (now Wisconsin Rapids). As it was distant and as some did not speak German, it was felt that there was a need for services in English. In 1907 services in English were offered in private homes. Eventually it was decided to build a church and to establish a definite connection with the Moravian Church in America. A church was built and dedicated in 1910.

With the immigration of numerous Catholics, especially of Poles, the Catholic Church developed extensively. Parishes were established, most of them for defined ethnic groups.

Father Dombrowski, who had started the church in Polonia, established a mission in the town of Hull and the church of St. Casimir was built in 1871. It became the second Polish parish in the county in 1874. Father Dombrowski also saw the need of Catholic schools. He obtained permission to have the nuns of the Order of St. Felix from Poland come to Polonia where they started a school in 1875. In 1877 the Generalate of the Order of Poland authorized the establishment of a province of the Congregation of the Felician Sisters in the United States with a motherhouse in Polonia. Since Polonia was too far distant from the large Polish parishes in the country, the motherhouse was moved to Detroit in 1882.

In Stockton, the first mass in Portage County was celebrated in 1840 in a private home by a Jesuit priest. Because of the large influx of Catholic Irish to the area, the Immaculate Conception of the Blessed Virgin Mary parish was established and a church built in 1875. It was also the Irish that built the church of St. Martin of Buena Vista around 1870 in the town of Almond, just south of the Buena Vista town line. The church was destroyed by a tornado and rebuilt in 1872 or 1873. It was because of the Irish too, that St. Patrick Church in Lanark was built in 1888.

Many churches were built to accommodate the large immigration of Polish people: St. Michael in Junction City and St. Bartholemew in Mill

Creek, 1883; St. Mary of Mt. Carmel in Fancher in the town of Stockton in 1884; St. Adalbert in the town of Alban near Rosholt in 1895; St. Bronislava just outside Plover in 1896 and, in 1897, St. Mary's for the congregation of St. Mary's of the Immaculate Conception in Torun.

In 1895 a church was built in Belmont. It was a wooden church which was replaced by a brick and stone building and named St. John the Baptist. In 1896 Bishop Messmer of the Green Bay Diocese (which then included Portage County) instructed parishioners that all Catholic children should attend Catholic schools and receive Catholic instruction. By 1873 Sisters of Notre Dame of Milwaukee had started the first Catholic school in Stevens Point. In 1975 the Felician Sisters opened a school in Polonia. The motherhouse of the Felician Province and the novitiate provided the teachers for the school in Polonia. When the motherhouse was moved to Detroit, a need was felt for a Polish order and teacher-training school in Wisconsin. After much controversy, especially with the Order of St. Francis, a German congregation, with some Polish nuns who had taken their vows, was formed under the name of the Sisters of St. Joseph. The order was approved by Rome and the motherhouse was built in 1902.

Two other orders of nuns have been active in Portage County. The Sisters of Notre Dame have their motherhouse in Milwaukee, but originated in Munich, Bavaria, Germany. These nuns have staffed St. Stephen School in Stevens Point since its opening in 1873. The other is the order of the Sisters of the Sorrowful Mother which originated in Bavaria and has a motherhouse in Milwaukee. When Stevens Point built a small county hospital, the Bishop of Green Bay suggested it be taken over by a religious community. In 1913 the Sisters opened St. Michael's Hospital.

Although the immigration of Europeans to America reached a peak between 1880 and 1890 and then declined, there was continuous immigration into Portage County, especially in Stevens Point where there were industries, businesses and education.

The main religious congregation continued as before: Methodists, Baptists, Episcopalians, Catholics, Lutherans and Presbyterians. The number of Catholics in Stevens Point increased greatly. These were people of Polish origin, largely descendants of pioneer farmers. A new Polish parish, St. Stanislaus, was organized and a church was built in 1916. Eventually the city was divided into four parishes by straight lines following streets. There would no longer be ethnic parishes, but parishes would be organized according to territorial boundaries. A number of wealthy families who were members of St. Stephen's Church and lived on Clark and nearby streets wished to remain in that parish. Certain blocks involving Clark, Main, Ellis, Reserve and Division streets were designated in St. Stephen parish and this extension maliciously was called the "gold thumb".

As farmers of Polish origin bought farms in regions originally settled by Yankees, Irish and Norwegians, there arose a need for new Catholic churches. In 1912, Guardian Angel Church in the village of Almond was dedicated. At first it was a mission of St. John the Baptist Church in Heffron, but later the roles were reversed and St. John the Baptist became a mission of Guardian Angel Church. In 1914, St. James parish was organized in Amherst. Services were first held in the old German Lutheran Church. In 1949, the cornerstone of the present church was laid. In Almond, the North German Methodists merged with the East German Methodists under the name of German Methodist Church. This church then merged with the English Methodist Church and became the United Methodist Church. In the towns of New Hope and Alban, the Lutheran Church of South New Hope was closed, but in 1923 a new church was built very close to South New Hope Church. It was Our Savior Lutheran Church affiliated to the Wisconsin Synod.

Religious congregations that at first had been isolated as ethnic groups speaking German, Norwegian, Polish etc., increasingly began to use the English language in their religious services. Ethnic parishes and churches had not accepted people of other ethnic groups as regular members, although in Catholic churches "foreigners" could attend Mass. The Church of the Immaculate Conception in Custer was at first the church of Irish farmers who had settled in the region. However Polish immigrants came to the area and although they could become sustaining members of the church, they could not be buried in the cemetery that was for "Irish only". Since the Irish were in majority in the township, they gained the majority of elective offices. But in 1898, the 18 sections of Stockton located north of the congressional township were turned over to Sharon. The Irish lost their political power, and in 1916 many moved to Stevens Point. As the Irish decreased in the Immaculate Conception Church, Polish families were admitted with full rights. But it was only in 1923 that the first Polish people were interred in the Custer cemetery.

Missionaries and newcomers introduced new denominations and churches in the Stevens Point metropolitan area: The Church of Christ Scientist formed in the 1870s following publication of "Science and Health with Key to the Scriptures" by Mary Baker Eddy, interested people in the Stevens Point area. In 1912, regular Sunday services were held in various homes. In 1926, the Mother Church in Boston accepted the group as a society and it became known as Christian Scientist Society, Stevens Point. In 1935 members purchased the former German Methodist Church and remodeled it in 1938. In that same year the society qualified as a branch church and became known as the First Church of Christ, Scientist, Stevens Point. In 1957, the present church was built.

The Jehovah's Witness congregation was established in Stevens Point in 1916. After meeting first in private homes, the worshippers rented the Odd Fellows Hall for their services in 1930. In 1943 they established Kingdom Hall and in 1957, erected the present building.

The Assembly of God Gospel Tabernacle congregation was established in Stevens Point in the early 1920s. At first it was an Assembly of God Mission which held meetings in a tent had out-of-town speakers. In 1931 the local church received a certificate of fellowship that the tabernacle had entered into the cooperative fellowship with the Wisconsin and Northern Michigan District Council of the Assemblies of God. A tabernacle was built in 1932 and in 1986, construction was begun for a new tabernacle.

The Church of Christ in Portage County is the result of a number of years of development. In 1894, an evangelist began preaching in Stevens Point, Mill Creek and Linwood. Small congregations were formed, and in 1944 when an evangelist took up residence in the area, the church was organized. In 1952, the group purchased a building and just a few years ago the present church was built on the outskirts of the city.

In recent years, new religious groups have met. Some have their own buildings: The Church of Jesus Christ of Latter-Day Saints (Mormons), Evangelical Free Church of Stevens Point, Berea Baptist Church, Grace Baptist Church, Greater Point Baptist Church, Divine Word Evangelical Lutheran Church, Good Shepherd Lutheran Church, Seventh-Day Adventist Church of Stevens Point, Apostolic Bible Chapel and Good News Fellowship.

Other groups meet in private homes or other locations: Religious Society of Friends (Quakers), Unitarian-Universalist group, Apostolic Bible Chapel, Pentecostal Church of God and Church of the Nazarene.

As the Stevens Point Normal School grew to become a campus of the University of Wisconsin with more than 9,000 students, a need for providing religious care for students and faculty became evident. Catholics organized the Newman Center on campus and in an oratory within the building, weekday Masses are held. Sunday Masses take place in St. Joseph Convent Chapel. Newman Center is also host for United Ministries in Higher Education, a Protestant campus ministry supported by American Baptist Church, Presbyterian Church, United Methodist Church and United Church of Christ. The Lutheran Church has its own building, the Peace Campus Center.

In rural areas, there have been a number of changes: South New Hope Lutheran Church has been reunited to North New Hope Lutheran congregation. Alban Lutheran Church and Concordia Church joined to form Faith Lutheran Church in Rosholt. Grant Methodist Church and Blaine Methodist Church, both in the town of Belmont, have closed and parishioners transferred to United Methodist Church of Almond. St. Martin parish in Ellis was closed and its territory divided between the parishes of Sacred Heart, Polonia and St. Mary's, Custer.

In Stevens Point, and in the county, several church buildings were damaged or destroyed by fire. Some were rebuilt in the same place; others were rebuilt at another location. Some churches moved to a new location. Formerly, churches were built to be near the center of the city; today churches are located where there is enough space for a large construction and abundant parking space.

In the United States, there are numerous branches of the Protestant sects. At present in Portage County, there is only one Methodist group, the United Methodist Church. Three Lutheran groups are represented: American Lutheran Church which has united most Lutheran groups of Scandinavian origin, the Lutheran Church of the Missouri Synod comprising most churches of German origin, and Wisconsin Evangelical Synod which includes churches which did not join the American Lutheran Church.

At first, Catholics of Portage County were in the Milwaukee Diocese. In 1868, with the creation of the Green Bay Diocese, the area east of the river was included. The western part of the county was attached to the La Crosse Diocese. In 1905, the Superior Diocese was established and the entire county became part of the Green Bay Diocese. In 1945, the Madison Diocese was created and Portage County was transferred to the Diocese of La Crosse, where it is today.

The Polish National Catholic Church which started in Scranton, Pennsylvania in 1897 was a movement which did not recognize the authority of the Pope. It has never had a parish in Portage County but there is a congregation in Mosinee, Marathon County, not far from the county line.

Baptists are divided into several groups but all other Protestant congregations are independent. In Portage County only one congregation has been dissolved: Beth Israel, the Jewish group. It is the Jewish tradition to hold services in Hebrew. Children attend classes to learn the language and the principles of their religion. The rabbi is the head who leads prayers and may teach the children. Since members of Beth Israel were decreasing, the last rabbi in Stevens Point was in 1966. For a time, one member who could speak Hebrew led services. On special holidays, especially Roshnashana (New Year) and Yom Kippur (Day of Atonement), a rabbi from Chicago came. According to the rule, at least ten confirmed male members must be present for a service to be held. As this number could no longer be reached, in 1985 the congregation was dissolved. The synagogue has provisionally been handed over to the Portage County Historical Society with the hope that it may become a museum.

The Amish who have recently settled in the county, represent a special study. Following their tradition, the Amish meet for religious service in family homes and use the German language. When there are too many families for one house, the community splits and a second community is formed, moving sometimes to another location. When the first Amish came to Portage County in 1966, the families formed one church district; now there are three districts and the number of families has increased to about 70.

Stevens Point has become the service center of the county. All schools teach English, which is now the language universally used even by persons of foreign ancestry. With the consolidation of schools, with military service, with men and women working in factories -- young people had the opportunity to meet people outside their ancestral group and marriages uniting individuals of different faiths, resulted. Children were educated in the religion of one of their parents, or were even attracted to another denomination. Consequently Polish names are no longer exclusively attached to Catholicism, and Scandinavian names no longer to Lutheranism, but can be found among the rosters of any church. The majority of Polish peoples, however, remain Catholic and most Norwegian, Lutheran.

There are more than 80 cemeteries in Portage County. Most towns and villages have their own, but some congregations have exclusive burying grounds: Catholics, Lutherans, Amish and Jehovah's Witnesses. Moravians in the town of Grant are buried in a cemetery in Wisconsin Rapids, although St. John Lutheran Cemetery in Keller is close to the Moravian Church.

In the rural areas, in early days, burials took place on the family farm or in a small cemetery set aside for people in the vicinity. Many of these cemeteries are abandoned; some are still kept up however, even though the last burial was 50 or 60 years ago. The Lombard Cemetery near Arnott in Stockton has graves of about 30 people, all with the name of Lombard. The earliest monument dates to 1868, the most recent, 1908. Although there is no longer anybody by the name of Lombard in the county, the cemetery is well kept. Cemeteries were often named after a family; the name usually found on the most prominent marker: Bickel, Mehne, Martin, Pratt. Some bear the name of the town, the parish, the congregation. Some have picturesque names: Oak Grove, Elmwood, Hillside.

Styles of tombstones reflect the prevalent tastes of people at the time. Until 1860, markers were slabs of white sandstone or limestone with inscriptions finely incised, often in italics to imitate handwriting. Each stone represented an individual: name, date of death, and date of birth. For children, the first name and middle initial was given, whether a son or a daughter, and the name of the father and mother.

During the 1860s, the style of stones became more elaborate. The top of the slab was rounded or pointed and symbols were often added: a weeping willow for sorrow; a crown for victory over death; a rose for virtue; clasped hands for friendship; a forefinger pointing heavenward, or a half-open gate for the entry into paradise. For young children: a lamb for innocence; a dove for gentleness, or an angel to carry the youngster to eternal life. The grave of a soldier or veteran showed a flag, and members of fraternal orders displayed their identifying marks: rings for the Odd Fellows; compass and trowel for the Freemasons. Also during the 1860s, epitaphs appeared, apparently introduced by German Protestants.

A number of changes were introduced in the 1870s and 1880s. The entire monument in many cases became symbolic: a broken column for a life that ended early; an obelisk signifying eternal life; an anchor for hope; a tree for regeneration. Individual graves began to be replaced by family plots, one monument being used for all members. This trend reduced the number of symbols and epitaphs used, and eventually they were discontinued altogether. Obelisks and columns increased in size: in 1900 some measuring 10 or 15 feet. Eventually slabs, obelisks and trees were replaced by massive blocks. Often the blocks were marked with the family name only, and small stones for each member of the family were aligned at the front or back of the plot. Red granite of Wisconsin (Wausau and Montello) was popular for monuments, but other materials were also used: grey granite, concrete, marble and iron. As fewer children died at a young age, family plots were replaced by a single block for a couple, bearing the names of husband and wife. In many cases, the stone was placed at the death of one of the spouses with the name and birthdate of the surviving mate already engraved. Only the date of death is left blank. In some cases, stones were already in place while the couple is still living. Occasionally, a photograph of the deceased is inlaid, or a picture of the couple, even though one is still living.

Cemeteries, especially the older areas of them, reflect the religion, origin and traditions of the early settlers. Catholics used a cross at the top of monuments. Protestants didn't use crosses, but often had an urn or an obelisk. Most inscriptions were in the native tongue of the deceased: German, Polish, Norwegian, Danish. With time, the inscription was in English but the Bible verse was in the native language. Even today, grave stones have inscriptions in Polish. Many tombs indicate the birthplace of the deceased: born in Norway, Germany, Ireland. Especially for the Irish, there was the addition of the name of the county and the parish. Inscriptions often represented old country customs. For the Polish language, the various genders of masculine, feminine, neuter. In St. Peter Cemetery in Stevens Point, there is a monument to the Lewandoskich family with the names of Jakob Lewandowski and Jozefina Lewandowska. Before coming to this country, many Norwegians had no family name. They had a first name, and the second name showed the filiation: Peterson, son of Peter, Petersdatter, daughter of Peter. The first son generally had the same first name as his grandfather. In the cemetery of North New Hope, is the family tomb with the following names:

NILS OG MRS MARTHA GRUNDESEN
GRUNDE NILSEN
MIKE NILSEN
MARTHE NILSDATTER
GINE NILSDATTER
KARL NILSEN

Cemeteries may also show settlers' attachment to their native country. In Nelsonville Cemetery a monument of red granite (similar to Wisconsin granite) bears the notice "Granite from Norway."

Burial sites of early pioneers are unknown. They may be in existing cemeteries but had only a wooden marker that disappeared or may be in unmarked graves in other locations.

It appears the earliest burial recorded by a tombstone is the one in Morrill Cemetery in the town of Plover. It reads:

LOUISA
dau of J.J. & S.J. Moor
died July 10 1846
aged 1 month 2 d

In the same cemetery there is an obelisk with on one side:

LAIA BLUE
Wife of
JAMES GEE
died
Feb. 12 1832
aged
35 yrs 5 months

but it is certainly only to record the name of James Gee's wife for she was born in 1798 and died in 1871 before coming to Portage County. The obelisk was erected at James Gee's death, his wife's name added at that time.

Crosses and Shrines

Visible signs of religion in the county are the numerous church buildings, especially those with towers and spires rising above houses and above trees, and seen from miles away. There are crosses and shrines near Catholic churches and cemeteries, but also in fields sometimes very distant from the churches. This custom was brought from Poland where farmers erected crosses or shrines near crossroads. Thus Polish farmers living far from villages would have a shrine at which to worship; or a cross to protect their crops from disaster; a place to give thanks for good crops or to stop and say a prayer on the way to church. Those shrines in Portage County have been erected by Polish people remembering their native country. In the summer, people bring flowers to the shrines. Some farmers have their own shrines; often it is a bathtub set on end in which a statue of the Virgin Mary or a saint is placed. The latest dated cross is from 1959, but in recent years some shrines and crosses have been taken away to make room for plowed fields.

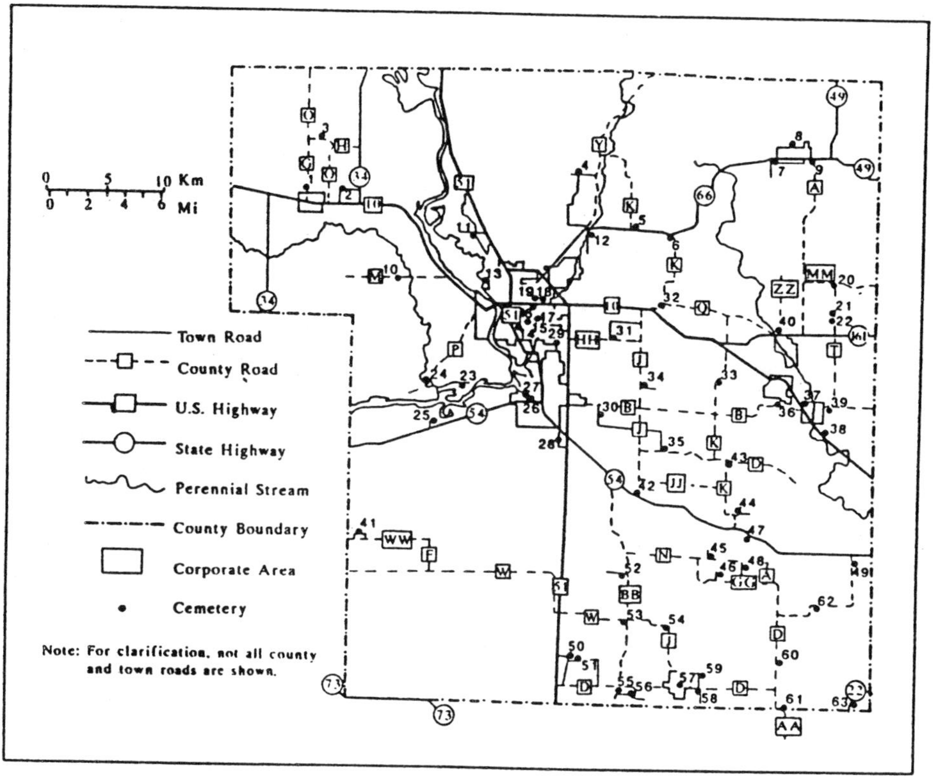

CEMETERIES
Source: Schulz

CEMETERIES IN PORTAGE COUNTY
MAP LEGEND

Cemetery	Town	Number on Map
1. Almond Village Cemetery	Almond	57
2. Amish Cemetery	Stockton	44
3. Arnott Cemetery	Stockton	34
4. Badger Cemetery	Lanark	49
5. Belmont Town Cemetery (Blaine-Kent Cemetery)	Belmont	60
6. Bickel Cemetery	Amherst	36
7. Buelow Lutheran Cemetery	Eau Pleine	2
8. Concordia Cemetery	Alban	8
9. Greenvale Cemetery	Belmont	63
10. East German Cemetery (German M.E. Cemetery)	Almond	58
11. Elmwood Cemetery	Belmont	62
12. Faith Luthern Cemetery	Alban	9
13. Forest Cemetery	Stevens Point	15
14. Ferman Evangelical Lutheran Cemetery	Stevens Point	17
15. Greenwood Catholic Cemetery	Amherst	37
16. Guardian Angel Cemetery	Stevens Point	18
17. Guardian Angel Catholic Cemetery	Almond	56
18. Hillside Cemetery	Lanark	48
19. Holtz Cemetery	Buena Vista	46
20. Jordan Cemetery	Hull	12
21. Kimball Cemetery	Buena Vista	43
22. Liberty Corners Cemetery (Buena Vista Cemetery)	Buena Vista	42
23. Linwood Cemetery	Linwood	23
24. Linwood-Union-Woodlawn Cemetery	Linwood	24
25. Lombard Cemetery	Stockton	35
26. Lone Pine Cemetery	Almond	53
27. Lower Amherst Cemetery (Lower-Fleming Cemetery)	Amherst	38
28. Maine Cemetery	Plover	28
29. Martin Cemetery (Mehne Cemetery)	Almond	54

30. McDill Cemetery	Plover	29
31. Meehan Cemetery	Plover	25
32. Morrill Cemetery	Stockton	30
33. Nelsonville Cemetery	Amherst	40
34. North Hope Lutheran Cemetery	New Hope	20
35. Oak Grove Cemetery	Amherst	39
36. Our Saviour Lutheran Cemetery	New Hope	21
37. Pine Grove-Bluff Cemetery	Pine Grove	50
38. Plover Cemetery	Plover	27
39. Pratt Cemetery	Pine Grove	51
40. Runkle Lutheran Cemetery	Eau Pleine	3
41. Sacred Heart Catholic Cemetery	Sharon	6
42. St. Adalbert Catholic Cemetery	Alban	7
43. St. Bartholomew Catholic Cemetery	Carson	10
44. St. Bronislava Catholic Cemetery	Plover	26
45. St. Casmir Cemetery	Hull	11
46. St. John Catholic Cemetery	Belmont	61
47. St. John German Lutheran Cemetery	Almond	59
48. St. John's Lutheran Cemetery	Grant	41
49. St. Joseph's Cemetery	Stevens Point	19
50. St. Martin Catholic Cemetery	Almond	52
51. St. Martin Catholic Cemetery	Ellis-Sharon	5
52. St. Mary Catholic Cemetery	Custer-Stockton	32
53. St. Mary Catholic Cemetery	Torun-Dewey	4
54. St. Mary Catholic Cemetery	Francher-Stockton	33
55. St. Michael Catholic Cemetery	Eau Pleine	1
56. St. Patrick Catholic Cemetery	Lanark	47
57. St. Paul Lutheran Cemetery	Buena Vista	45
58. St. Peter's Cemetery	Hull	13
59. St. Stephen's Cemetery	Stevens Point	16
60. South New Hope Cemetery	New Hope	22
61. Spiritland Cemetery	Almond	55
62. Stockton Cemetery	Stockton	31
63. Union Cemetery	Stevens Point	14

HEALTH

As early as the year 1845, a doctor resided in Stevens Point. When he died in 1848, a young graduate in medicine took his place.

The 1850 Portage County census lists two doctors, both in Stevens Point. As the decade progressed, the county attracted more doctors, most of them recent graduates of medical school. The 1893 city directory listed 11 physicians and surgeons. Doctors were located in some of the other cities also: Almond, Amherst, Arnott, Junction City, Nelsonville and Rosholt. In Nelsonville a doctor built a two-story house to be used as a hospital, office and drugstore, but his venture failed and he moved away. In 1899 a man born in Oslo, Norway in 1872 came with his parents to Chicago where he attended medical school. He graduated in 1896 and interned at the Norwegian Hospital in Chicago where he heard that Norwegians in Portage County were in need of a doctor. The young doctor came to Nelsonville and practiced there until he died from a heart attack in 1934.

In 1962 there were 20 physicians and surgeons in Stevens Point, one physician in Amherst, one in Rosholt and one in Whiting. An osteopath, a podiatrist and two chiropractors were also practicing in Stevens Point. In 1989 there were about 60 physicians and surgeons in the Stevens Point area, mostly specialists in the fields of dermatology, gynecology, internal medicine, neurology, obstetrics, opthamology, orthopedics, psychiatry, surgery and urology. There is no longer an osteopath.

Most physicians are working in groups or clinics; the oldest is the Rice Clinic but there are also others: Medical Arts Plaza, Doctors' Park, the Plover Medical Center and St. Michael's Hospital where doctors may refer their patients.

There were several private hospitals in the 19th century. The Stevens Point Women's Club was the impetus behind a city hospital in Stevens Point. Construction began in 1911 and the first patient was admitted on January 23, 1913. The Sisters of the Sorrowful Mother became administrators of the city hospital, St. Michael's. Since its opening, St. Michael's has undergone several expansions. In 1976 the staff was made up of 45 physicians and 120 nurses who worked full time. In addition there were volunteers who aided the nurses. Outside of Stevens Point there is only one physician in Rosholt.

There are about 12 chiropractors in Stevens Point and one in Amherst.

In 1936, Ruth Gilfrey, a nurse, initiated a nursing service, which in 1976 became the Health Department. At that time there were eight nurses, two sanitarians, one health aid, one planning and three clerical personnel under the jurisdiction of the County Health Officer. Over time the nursing service under the umbrella of the Health Department developed and performed various functions such as counseling, school health services, an immunization clinic, a venereal diseases clinic, a blood pressure control clinic and family planning. Eventually the Health Department took over functions held by the Environmental Health Department, including inspection of public facilities according to sanitary codes (parks, restaurants, taverns, hotels and motels), controlling industrial safety, inspecting food in stores and testing the quality of water.

In 1978 a new building was dedicated and named the Ruth Gilfrey Human Resources Center. In it are housed various social services. Once a month the Health Department provides family clinics in Almond, Amherst, Bancroft, Junction City and Rosholt.

The County Ambulance Service was first established in 1947. In 1976 four ambulances were stationed at the Fire Department. Since 1965 there has been pressure for increased ambulance service and staffing more paramedics, highly skilled emergency personnel who are trained to provide all necessary first aid.

For individuals who suffer incurable diseases, the elderly, the disabled and those persons who have no one to take care of them, there are two homes, both surrounded by large wooded parks: Portage County Home and River Pines Living Center. The first, a government institution, was established in 1964 and known as Portage County Infirmary. The latter, a private institution founded by two doctors in 1906, began as a sanatorium for tuberculosis patients. In 1938 it was taken over by the Sisters of St. Joseph of the Third Order of St. Francis. As a sanatorium, it had had up to 60 tuberculosis patients, but because of advances in the diagnosis and treatment of the illness, the number of patients has dwindled to about a dozen. The sisters sold it and the buyer built a 238 bed nursing home. In 1974, when River Pines Community Health Center became part of Mid-American Health Service operated by the Seventh-Day Adventist Church, the name was changed to River Pines Living Center.

In early times, physicians prepared medications for their patients. It was not until 1821 that the first American College of Pharmacy was founded in Connecticut.

Looked upon as an American institution, drugstores were the only businesses authorized to be open on Sundays. In addition to drugs, the stores added other articles little by little. First it was the Sunday newspaper, then tobacco, cigarettes, sweets, greeting cards and cosmetics. Some drug stores added a counter where they offered ice cream, soft drinks, coffee and sandwiches. Some even added booths with tables. But there was always a pharmacy section with one or two pharmacists, although in some cases this department was overshadowed by other sections of the store.

Although it was not known when the first drug store or pharmacy in the county was opened, the Stevens Point Directory of 1889/1890 listed four druggists. The directory of 1896 had six drugstores, the same number found in the 1962 directory. They were operated by the owner, but in the 1960s and the 1970s, new drugstores opened as part of store chains within a medical clinic. Now, there are ten pharmacies in the Stevens Point area, but only two of them are operated by the owner. Gwidt Drug Store, owned and operated by Mr. Leo Gwidt since 1945, closed in 1990. It was founded in 1919.

Dental care in Stevens Point started after the Civil War when two dentists came to the city to begin practice. In 1889 there were three dentists in the city, ten in 1896, eight in 1915, five in 1931, 14 in 1952 and 24 in 1976. Two orthodontists and one oral surgeon practiced out of their office in the Wilshire Professional Center in 1976. At this writing, there are approximately 30 dentists in Stevens Point.

In 1989 the Stevens Point City Directory listed one optician; in 1896, one oculist and aurist (eye and ear) and, in 1927, there were two opticians and two optometrists. Now there are ten optometrists (O.D.), five of which are connected with clinic or optical businesses. There is no optician listed. There are three optical businesses in Stevens Point; the same is true of Amherst.

In the field of mental health, there are several doctors (M.D., Ph.D, Ed.D, Psy.D, D.SW.) and social workers (M.SW., A.C.S.W.). They offer their services at Ruth Gilfrey Community Human Services, St. Michael's Hospital, Rice Clinic, Medical Plaza, Riverdale Clinic and the Mental Health Clinic of Central Wisconsin. All buildings are located in Stevens Point.

In early times when a pioneer died, his companions buried him in a coffin made by a logger or lumberman. Later, building caskets became the task of carpenters and cabinetmakers. They were the ancestors of undertakers who eventually took the name of funeral directors. Because, at first, they could not make a living from funerals, they started furniture stores. The Boston Funeral Home was the first in Stevens Point, 1888. In 1890 there were three undertakers: Boston, Grant and Kamrowski, each with a funeral home under his name, as well as a furniture store. In 1903 there were three undertakers listed: Boston, Kelly and Rosenow, the last two falling under the heading "embalmers." In 1915 the three undertakers listed were Boston, Kamrowski and Rosenow. In 1931 four funeral directors were listed: Boston, Prais, Rosenow and Kamrowski, the first three also dealt in furniture. In 1950 five funeral businesses were listed: Boston, Crosby, Prais, Rosenow and Dzikoski. In 1954 there were five: Boston, Crosby, Prais, Rosenow and Dzikoski. The last director, who was Polish and located across the street from St. Peter Church (the Polish Catholic Church), handled most of the Polish funerals. In 1973 there were four funeral homes: Boston, Crosby, Dzikoski and Shuda. Boston had about 90 funerals per year. They had handled up to 150 funerals a year in 1940, before Dzikoski started a funeral home of his own. Dzikoski handled approximately 150 funerals in 1973, Crosby had 40 and Shuda had about 50.

There were also funeral homes in Almond, Amherst and Junction City. Now, there are only three funeral homes in Stevens Point: Boston, Dzikoski and Shuda. The funeral home in Almond closed because the owner also operated a home in Plainfield. There is a new funeral director in Plover.

The wake, which was originally held in the home of the deceased, is now generally held in the funeral home with an open casket. The corpse is embalmed; enabling relatives and friends to retain a pleasant memory of the deceased. Embalming is also necessary because of climate. In winter the cemeteries are closed because the ground is frozen. Caskets must be kept in refrigerated rooms of funeral homes awaiting spring for burial.

Health care for diseased animals by specialists began rather late. In the 1860 census there were no veterinarians. But in the directory of 1889, there were five, in 1893 the number declined to three, in 1903 there were only two and from that time on there were only one or two listed. In the 1960s two doctors of veterinary medicine opened hospitals for small animals. Now, there are 14 veterinarians total. There are four hospitals with eight doctors, and two veterinary services for large animals.

POSTAL SERVICE

The first post office in Portage County was opened in 1845 at Plover. Fifty-three post offices followed.

Initially, there were no home mail deliveries. People had to go to the post office to claim or mail letters and packages. Consequently, post offices were opened soon after immigrants settled in an area.

In 1849 post offices were opened in Stevens Point and Shawrette; in Almond and Buena Vista, 1850; in DuBay (the name was changed to Eau Pleine), 1851; in Lake Sarah, 1852; in Iron Creek, 1854; in Badger and Madely, 1855 and in Emily and Stockton, 1858. Some were short-lived: Shawrette closed in 1850, Lake Sarah in 1854 and Iron Creek in 1857.

In 1900 there were 35 post offices, but their number decreased rapidly with the establishment of rural routes in 1896. Between 1900 and 1907 21 post offices were closed, including one at Torun that lasted from 1902 to 1904. Only one post office opened after that at Coddington, but it closed in 1955. The one at Tabor opened in 1907, but closed in 1912. Casimir closed in 1910, Fancher in 1913, Arnott and Stockton in 1934 and Polonia in 1939.

At present there are only ten post offices in the county: the city of Stevens Point; the villages of Almond, Amherst, Amherst Junction, Junction City, Nelsonville, Plover and Rosholt and the unincorporated places of Bancroft and Custer. Since the Nelsonville post office is so near Amherst Junction, it has been an issue to close the post office. However, it remains open since inhabitants of the village, many who are elderly, have petitioned to keep it open. With the exception of Nelsonville, the other post offices have free rural delivery.

Post Offices

(1845 - 1870)

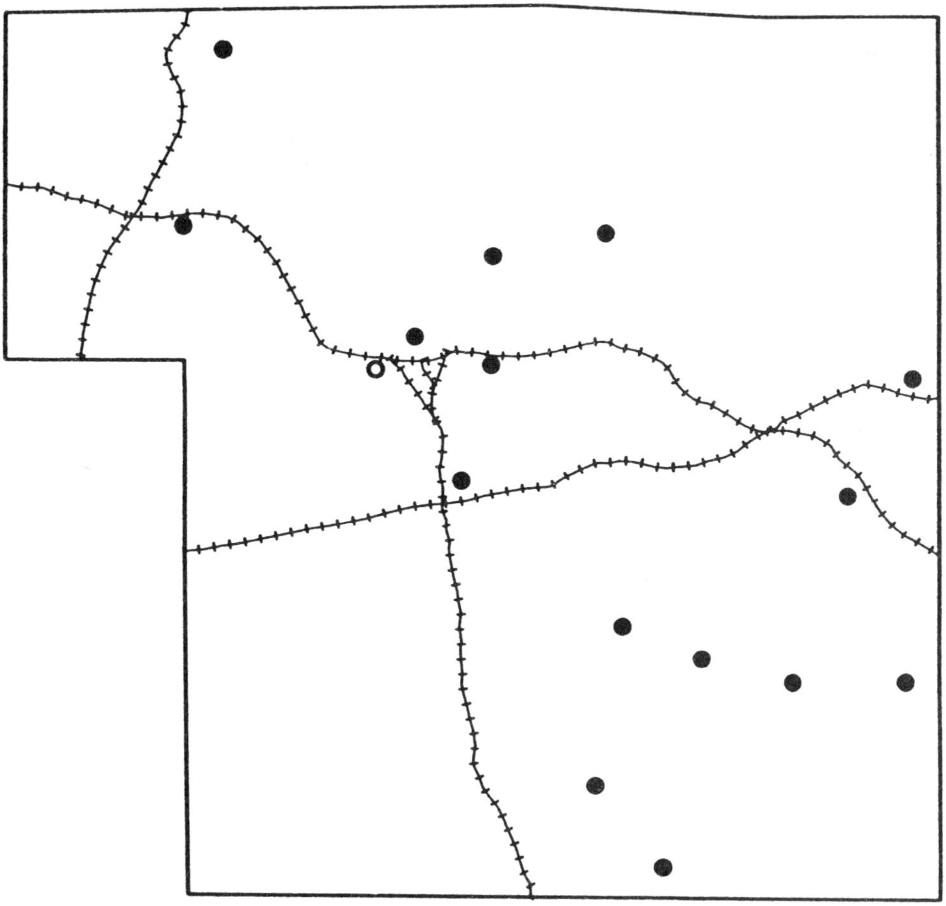

Open in 1870. . . . . ●

Closed between 1845 and 1870 . . . . o

Railroad. . . . . ++++++

Post Offices

(1870 - 1900)

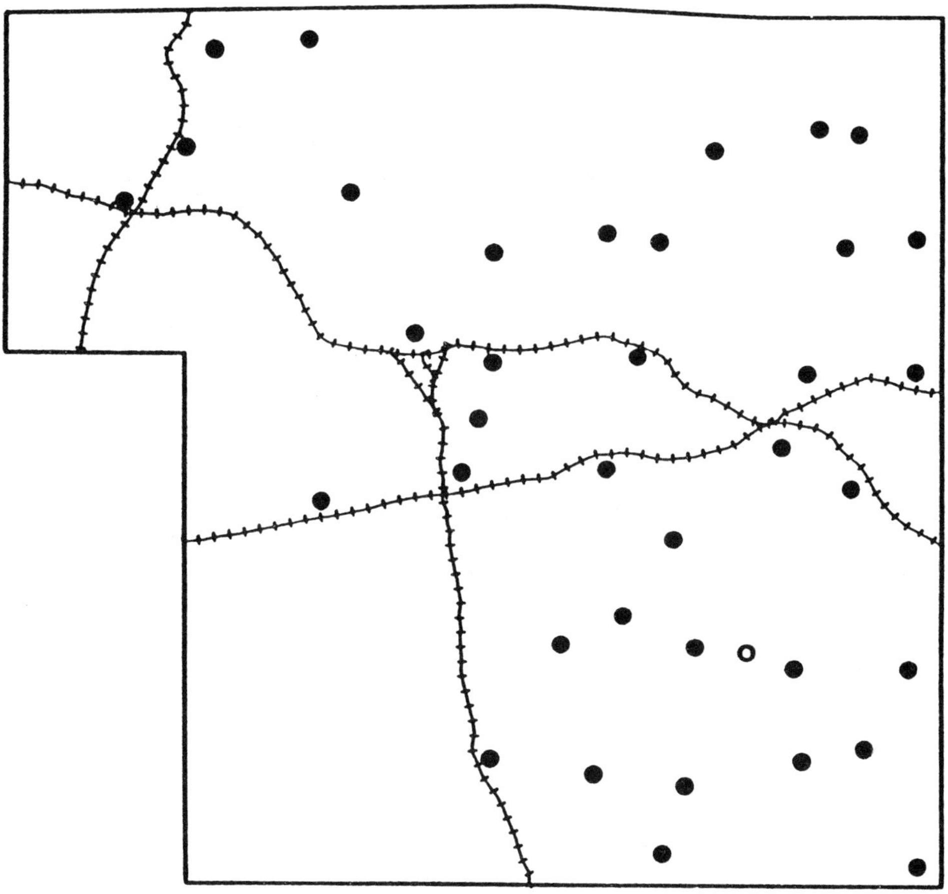

Open in 1900. . . . . ●

Closed between 1870 and 1900 . . . . o

Railroad. . . . . +++++++

Post Offices

(1900 - Present)

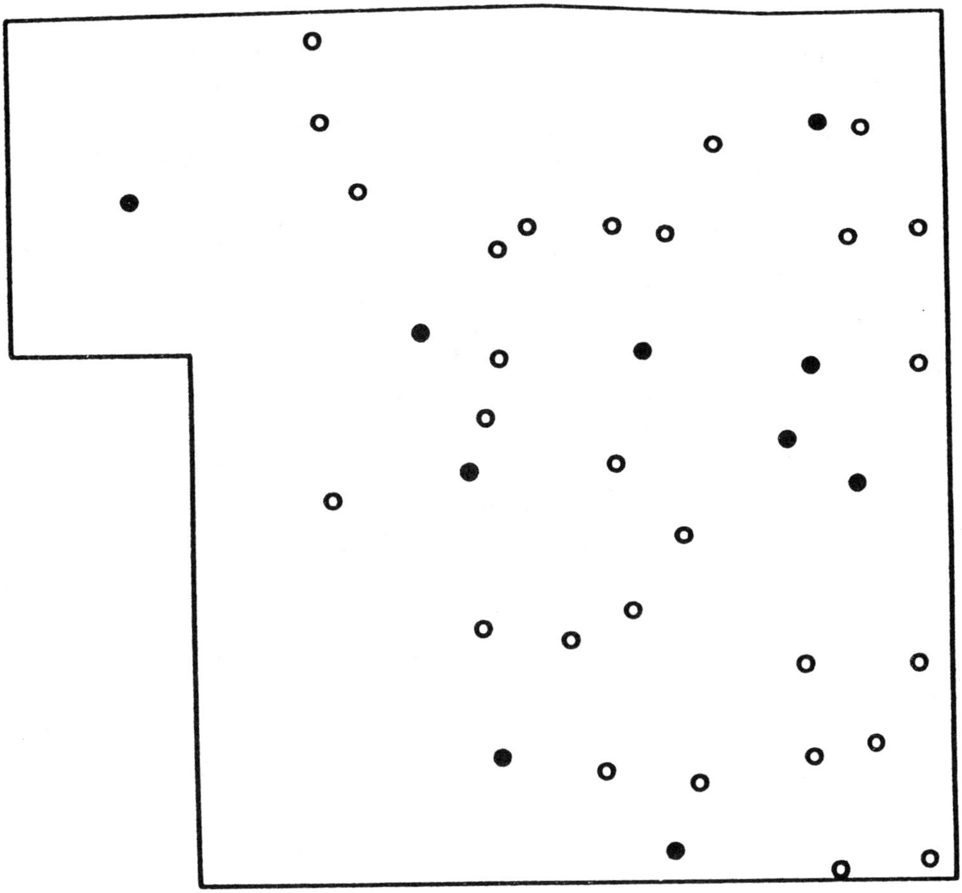

Presently open. . . . . ●
Closed between 1900 and present . . . . . . o

NEWSPAPERS

Since Stevens Point was considered the most important village in the county, it is no surprise that it had the first newspaper in the county. A weekly, the Wisconsin Pinery, was first published January 14, 1853.

Plover, the county seat, also had a weekly newspaper, the Plover Herald, first published in 1856. From 1857 to 1861 the newspaper was published under the name of Stanton Times, the name some of the inhabitants preferred to call their village. In 1864 the newspaper was consolidated with the Republican and its name became the Stanton Times and Republican. In 1863 the name was changed again and it became the Plover Times. Although the county seat was moved from Plover to Stevens Point in 1867, the Plover Times continued to be published until 1881. Since that time however, no newspaper has been published in Plover.

In 1863 another newspaper began publication in Stevens Point. It was called The Wisconsin Lumberman, but in 1870 the name was changed to Point. In 1872 the newspaper was renamed to Stevens Point Journal.

In 1878 the Portage County Gazette was founded as a weekly. It became a daily in 1885. In 1919 the Gazette was consolidated with the Stevens Point Journal. Because the latter newspaper had become a daily publication in 1895, the consolidated version took the title of Stevens Point Daily Journal. It is presently the only English language newspaper published in Stevens Point. Gwiazda Polarna (Northern Star), a Polish language weekly, was founded in 1908 and continues publication to this day.

A few additional newspapers were started in Stevens Point, but ceased publication after several issues or several years. Wisconsin State Rights, a weekly, came out in 1861. It was founded in 1859 in Monroe by N.L. Stout, who moved it to Stevens Point in March of 1861. Stout suspended publication in November of that same year when he enlisted in the Union Army.

In 1873 a monthly called Sensation began but lasted only until the following year. Hungerford's Real Estate Journal was published from 1879 to 1886. The Stevens Point Democrat, a weekly, was published from 1880 to 1885. Wisconsin Eagle, also a weekly, was founded in 1884 but ceased publication in November 1885.

In 1894 publication of a Wisconsin State Press weekly began. The same year the title was changed to The Stevens Point Press, which continued until 1896. From 1935 to 1939 Central Wisconsin Herald, a weekly, was published.

Newspapers in foreign languages also appeared but they remained for only a brief time except for Gwiazda Polarna. Niedziela (Sunday in Polish) was founded in 1872, but only lasted until 1874. From 1892 to 1899 the Stevens Point Post, a German language weekly, was published. Rolnik (Plowman), in 1891, lasted only a brief time. From 1930 to 1935, Jaskolka (Swallow), a monthly in Polish was published; from 1938 to 1940 Sloneczko (A Little Sun), a weekly Polish magazine for children, was published.

Villages also had their own newspapers but none has survived. Almond's Portage County Press, founded in 1901, became the Almond Press in 1924 but was discontinued in 1931. Amherst had the Amherst Advocate from 1893 to 1964, the Amherst Pioneer and the Tomorrow River Times from 1975 to 1981. Bancroft had the Bancroft News from 1908 to 1909, and Junction City had the Community Press in 1946. Rosholt had the Rosholt Echo from 1905 to 1906, the Rosholt Record from 1910 to 1913, the Rosholt Journal from 1913 to 1918, the Rosholt Review from 1920 to 1924 and the Community Press from 1926 to 1930. All village newspapers were weeklies.

Most newspapers were started as one-man projects with the purpose of expressing the publisher's ideas and opinions. Many newspapers died out when the first publishers quit and found no one to replace them.

TELEPHONE

Alexander Graham Bell invented the telephone in 1876. Only three years later, 1879, the first telephone was introduced to Stevens Point with a line connecting the Wisconsin Central Railroad downtown office with the freight depot. In 1882 the first exchange was installed in Stevens Point, and nine months later there were 79 subscribers.

Wisconsin Rapids, however, preceded Stevens Point with the first local telephone line in Central Wisconsin. The Bell Telephone Company established this line in 1878, and built its first toll line from Wisconsin Rapids in the spring of 1887. This single line connected the switchboard of the Wisconsin Telephone Company in Stevens Point with a single telephone located in a drugstore in Wisconsin Rapids. The idea of a new, locally-owned company originated at a meeting of the Wisconsin Rapids Businessmen Association. It was organized under a cooperative plan and, in 1896, the charter of the Wood County Telephone Company was granted. Service was established with a total of 119 telephones. Until 1907 there were no connections to the toll line, so that to make or receive a call from Stevens Point one had to go to the Bell Company.

In 1895 the Wisconsin Bell Company built a line through Amherst to the town doctor. The Amherst Telephone Company began in 1903, and the only telephone service available was a few lines that had been strung up in the village. The following year a few more lines were erected in the village and 14 miles (22.5 km) of rural lines were put up. The Amherst Telephone Company negotiated with the Wisconsin Bell Telephone Company, from 1904 to 1908, to take over service of this line. When this was accomplished, the Amherst Telephone Company had control of all telephone lines in the immediate area. In a 1925 decision to expand, the Amherst Telephone Company purchased the Nelsonville Farmers Telephone Company. This nearly doubled the size of the firm in both subscribers and miles of wire, since the villages of Nelsonville and Rosholt were now included.

In the 1890's the Almond Telephone Company was established. The Bell Company originally had a patent on the service, but the patent expired in 1893. Like mushrooms, independent companies sprang up everywhere in the county, often competing with one another. The greatest factor in destroying competition was the entry of regulation into the telephone field. In 1907 Wisconsin was one of the pioneers in state regulation of public utilities. The Railroad Commission of Wisconsin was given the power to regulate telephone companies. The Wisconsin Telephone Company, a subsidiary of the Bell Company, located its central offices in the largest towns and cities. Independents were located in the rural districts and smaller towns.

In 1926, the 50th anniversary of the telephone, Wisconsin Bell had 91 exchanges serving 299,068 subscribers. In the independent field, 964 companies operated more than 600 exchanges serving 184,326 subscribers. The Wisconsin State Telephone Association and the Northwest Telephone Company (Northwest Telecommunications), with their seat in Madison, have each consolidated many independent companies.

As of January 1, 1988, the Public Service Commission of Wisconsin counted 194 telephone companies and 546 exchanges, of which 96 depended on Wisconsin Bell, Inc. (including Madison and Milwaukee), and 192 on GTE North Wisconsin (including Wausau and Marshfield).

In Portage County, only one company has its own seat, the Amherst Telephone Company with three exchanges in Amherst, Rosholt and Polonia. Three other exchanges located in the county depend on companies having their seats outside the county. Stevens Point depends on Wisconsin Bell and Almond depends on Union Telephone Company located in Plainfield (which, in 1920, bought the Bancroft exchange). The Union Company has four exchanges: Almond, Coloma, Hancock and Plainfield. Junction City depends on Central State Telephone company of Vesper, which started in 1897. It has six exchanges: Auburndale, City Point, Junction City, Lindsey, Necedah and Pittsville. City Point is the smallest exchange in Wisconsin having 96 subscribers; its calling area counts 1,171 subscribers since it comprises of only City Point and Pittsville.

Some telephone subscribers depend on exchanges outside the county: Rudolph and Wisconsin Rapids are exchanges of the Wood County Telephone Company, Plainfield and Wild Rose are exchanges of North West Company, Waupaca is an exchange of Wisconsin Bell, Scandinavia and Iola are the only exchanges of the Scandinavia Telephone Company and Mosinee is the sole exchange of the Mosinee Telephone Company.

The number of subscribers of the exchanges and the number of callers within the calling area are as follows:

	Subscribers	Calling area
Almond	792	1,856
Amherst	1,460	3,158

Junction City	1,031	2,442
Polonia	907	21,361
(Stevens Point, Amherst, Rosholt)		
Rosholt	791	3,158
Stevens Point	18,203	19,110 (only Polonia)
	23,184	

Adding subscribers living in the county, but depending on exchanges located outside, and deducting subscribers to Portage County exchanges, but living outside the limits, one may count around 25,000 telephone subscribers in 1988 in the county.

FIRE DEPARTMENTS

Portage County's first fire fighting organization was founded in Stevens Point sometime around 1850. Known as the bucket brigade, the group consisted of a row of volunteers who passed buckets from a water source to the fire, and another row of volunteers who passed the empty buckets back. Because, like most early villages in Wisconsin, Stevens Point was built almost entirely of wood, it experienced a number of disastrous fires during the first decades of its growth.

The bucket brigade was given its first modern fire fighting equipment in 1860. This "hand-pumper" was first acquired by the Chicago fire fighters where it was used until retired from service. The pumps had long handles that were pushed up and down to build pressure.

As the volunteers became more aware of their responsibilities, they adopted a constitution and bylaws. It was looked upon as an honor and a mark of social distinction to serve with the volunteer fire fighters.

In January 1874, the most disastrous fire to date occurred. The fire destroyed many buildings around the Square, a block north to College Avenue and east along Main Street. As a consequence, citizens petitioned the council to purchase a steam engine. The engine was purchased in June of that same year. Despite the acquisition, there were several fires, and in 1885 another steam engine was acquired. In 1889, the two steam engines were retired and replaced by the use of hydrant pressure.

In 1892 the council adopted an ordinance to establish a paid fire department. In 1901 a fire alarm system was created, and 24 fire alarm boxes were set up in the city. In 1917 the first motorized fire vehicle was purchased. At this writing, there are 12 fire departments in the county. Stevens Point is the only one with 37 paid employees. There are four fire trucks--the oldest dating from 1967, the newest from 1988--all equipped with pumps and ladders.

In 1988 the Stevens Point Fire Department responded to 406 incidents, the highest number since 1984. Total loss from the 1988 fires was $113,543, or approximately $25,000 less than in 1987. The estimated loss per fire, based on Stevens Point's population of 23,000, was $4.93 per capita. In 1977, when fires destroyed nine businesses, damage totaled $529,780 and the per capita loss was $13.52.

The eleven other fire departments in the county are made up of volunteers. Each has its district, but in the event of a major fire, districts offer aid to neighboring districts. The districts vary in size and equipment.

The city of Stevens Point and the villages of Amherst, Amherst Junction, Park Ridge and Whiting have jurisdiction only within their boundaries. The fire department of Almond covers the village and towns of Almond, as well as the town of Belmont. The department of Bancroft (not an incorporated village) has jurisdiction in the town of Pine Grove and part of the town of Buena Vista. The department of Hull covers the towns of Hull and Dewey. Junction City's district, in addition to the village, covers Carson, Eau Pleine and Linwood in the county; the town and village of Milladore in Wood County and Bergen and Knowlton in Marathon County.

Nelsonville Fire Department is responsible for the village, the town of Amherst and part of the town of Lanark. The department of Plover answers fire calls from the village and town of Plover, the town of Stockton and the northern part of Grant. The department of Rosholt covers the village of Rosholt and the towns of Alban and Sharon.

Volunteer fire departments have been organized at various times. The oldest volunteer group is believed to date to the early 1900s in Amherst. The youngest is the Bancroft Fire Department, organized in 1989.

Besides Portage County volunteer fire departments, departments in other counties cover some areas of Portage County. The Iola Fire Department in Waupaca County responds to fires in New Hope, and the Wisconsin Rapids Fire Department to the western part of the town of Grant. The Waupaca Fire Department is responsible for the eastern part of the town of Lanark.

Volunteer fire departments number from 20 to 40 members who, in the event of fire, can be reached by telephone. All departments hold regular meetings. Men are trained through a Wisconsin vocational training program which may include setting an uninhabited house afire that, for some reason, had been previously marked for destruction.

Equipment of volunteer fire fighters varies from two to nine trucks--some are pumpers, some are tankers and several combine pumper and tanker. Fire departments receive finances from their own towns and from retainer fees paid by the other towns they serve. Some fire departments raise money through dances and picnics.

The average number of fires per year varies from district to district. Amherst averages only about five fires per year. Nelsonville averages between 12 and 15 per year; Almond about 20; Junction City and Plover 40 to 50--however the latter are mostly grass fires during dry periods.

Fighting forest fires is the responsibility of the Department of Natural Resources. Forest fires are considered to be those uncontrolled fires in forests, marshes or fields outside the cities or incorporated villages. The Department of Natural Resources is divided into six districts. Portage County is in the North Central District with headquarters in Rhinelander. The ranger station in Whiting is well equipped. Some years, during the fire season, rangers are called to control more than 100 fires. The fire season lasts from March to November, peaking in March, April and May. During these months, drought and high winds create high risk conditions for forest fires.

FINANCIAL INSTITUTIONS

Financial institutions can be divided into four groups: banks, savings and loan associations, credit unions and finance companies.

Banks may be either national or state. National banks have a national charter and are members of the Federal Reserve Bank, insured by FDIC (Federal Deposit Insurance Corporation). Many banks offer full service: checking accounts, savings accounts, savings certificates, safety deposit boxes, individual retirement accounts (IRAs), trust services and a variety of loans. The types of loans include: farm, business, consumer loans and mortgages. Banks may have drive-in windows, instant banking services available day and night (TYME) and other services for customers.

The first bank in Stevens Point was the First National Bank, established in 1883, with capital stock of $50,000. On the bank's 100th anniversary, August 1, 1983, the statement of condition showed stock at $900,000 and total liabilities and capital at $427,019. There is now a branch in Plover.

The Citizens National Bank of Stevens Point was founded in 1893. With time it became a full service bank. In 1986 it was acquired by the Marine National Bank of Milwaukee, which two years later sold it to Bank One of Columbus, Ohio.

In the villages the oldest bank is in Amherst, founded in 1893. With insufficient funds to obtain a national or state charter, the Amherst bank remained private and took the name of International Bank. Later, the bank became state-charterd and insured by FDIC.

The Bank of Portage County in Almond opened in 1901. In the 1970s, the bank became affiliated with Marshall & Ilsley and took the name M & I Portage County Bank. However, the bank became independent again under the name of the Portage County Bank.

In 1903 the Security State Bank of Amherst Junction was established. In the 1980's a branch of this bank was opened in Custer. In 1987 the Amherst Junction bank was purchased by F & M (Farmers and Merchants) Bank of Athens, but the name did not change.

The State Bank of Rosholt was established in 1907. It opened branches in Polonia, Elderon (Marathon County) and, in 1985, Stevens Point. In 1912 the Bancroft State Bank opened, and later established a branch in Kellner.

For 57 years, no new bank was established. Then the Bank of Plover opened in 1969, and the Bank of Park Ridge in 1975. In the 1980's both banks became affiliated with Marshall & Ilsley and added M & I before their names. However, the Bank of Park Ridge became independent again and dropped the M & I.

Another type of financial institution is the savings and loan association. There are only two such associations in the county: the American Savings and Loan established in 1924 and chartered by the state, and the First Financial Savings and Loan begun in 1965. The first was called Citizens Savings and Loan, and was affiliated with Citizens National Bank next door. Gradually Citizens Savings and Loan grew away from Citizens National Bank. In 1978 they moved to a new building and changed their name to American Savings and Loan.

The second company, First Financial Savings and Loan, has experienced a large expansion since its founding. It merged with the First Savings which had been in existence since 1934. At first it was known as First Federal Savings and Loan Association of Stevens Point. In 1973 it converted to a state charter as First Financial Savings and Loan Association. In 1980 it converted from a mutual to a state stock charter. In 1984 it merged with First State Savings of Waukesha creating the First Financial Corporation, one of the first multiple savings and loan holding company. In 1985 it acquired the First Savings Association of Wisconsin. In January 1989 it converted to a federal savings bank charter, the First Financial Bank, FSB. First Financial is insured by FSLIC (Federal Savings & Loan Insurance Corporation). First Financial counts among the largest savings institutions with 60 branches, one in Plover.

Previously, savings and loan institutions were very different from banks. They could not accept deposits and could not have checking accounts, but since deregulation, savings and loans may offer the same services as banks. Their clients, however, are different. Savings and loan associations lend money mostly for mortgages, but also for education, auto or other major purchases. Banks lend money to industry, trade, and agriculture.

The third type of financial institution is the credit union. The credit union is made up of employees and families who open savings accounts in the credit union to make low cost loans to other members who have purchased at least one share. Law requires that all members of a credit union have some common bond, usually occupational. The loans are small and short-term. After expenses are paid, any excess is distributed to members as dividends according to the number of shares they hold. The credit union is run by a board of directors who receive no monetary reimbursement, and are elected by members. Credit unions are the only democratically controlled financial institutions. Only members may vote.

The first credit union was organized in Germany, in 1849, by Mayor Fredrich Wilhelm Raffeisen. At that time a depression had crushed townfolk and farmers alike, so Raffeisen rallied villagers to pool their savings and lend to each other at low interest.

A Canadian reporter, Alphonse Desjardins, founded a credit union in Levis, Quebec, in 1900. In 1909 the same man founded the first credit union in the United States at Ste Marie, New Hampshire. Today there are approximately 15,000 credit unions in the country and more than 57 million members.

There are seven credit unions in Portage County, all located in the Stevens Point area. The oldest, organized in 1935 for the employees of Consolidated Papers, was the Consolidated Credit Union. Later it became the credit union for the employees of Stevens Point Concrete Company and the Catholic schools, as well.

In 1948 the Wisconsin River Credit Union, for employees of the Wisconsin River Division of Consolidated Papers, was formed. In 1953 the Whiting Plover Credit Union opened, the only residential credit union accepting residents of Whiting and Plover who are employed but ineligible in their credit unions.

In 1954 the Point Soo Credit Union was organized for employees of the Soo Railroad. Today however, the union accepts employees of more than a dozen firms, the largest being Copps Corporation and Herrschners. In 1964 the Co-Ci Credit Union was formed for the employees of the city of Stevens Point and of Portage County. In 1987 it became a branch of Mosinee Credit Union. In addition to city and county employees, the credit union served members of the Board of Education, employees of Mid-State Technical Institute, the Rice Clinic and Community Industries. In 1967 the University of Wisconsin-Stevens Point Credit Union was formed and, in 1981, the Sentry Credit Union.

Bull's Eye Credit Union in Wisconsin Rapids has a charter including the city of Stevens Point, the villages of Plover and Whiting, the towns of Buena Vista, Carson, Grant, Hull, Linwood, Plover and Stockton.

In addition to the three major financial groups, there are more than 20 companies or corporations dealing with finances: planning, investments, mortgages, stocks and bonds. Some are associated with insurance and real estate companies; others are agents of finance companies located outside the county or outside Wisconsin.

INSURANCE

In 1887, the New Hope Norwegian Mutual Insurance Company was founded. Its bylaws stipulated that at least two-thirds of the company officers be Norwegians. An article in the bylaws required that the headquarters be located in New Hope and so it was--in the hamlet of Peru in New Hope.

At the second annual meeting, the board of directors decided to insure Americans and Germans as well. The company was formed to write fire, wind, hail, lightning, theft, vandalism and liability insurance on farm and city property in Portage, Waupaca and Marathon counties. During the first year of its existence, the company had no claims against it, and in 1891 losses amounted to a mere $12.67. At the present writing, the company has approximately 1,300 policy holders. In 1985, collections reached nearly $287,000 in premiums; paid-out claims amounted to less than half that amount. By 1935, the name Norwegian was dropped from the company name.

Portage County is also the headquarters of a giant insurance company, Sentry Insurance. Sentry's humble beginnings centered around the concerns of a handful of hardware dealers in Wisconsin who felt that insurance companies charged too much. These dealers had flammables in their stores but few losses by fire, so they agreed to band together to work for their common interests in what was eventually named the Wisconsin Retail Hardware Association.

In 1903, the association formed a mutual insurance company furnishing fire insurance at a much lower cost than other companies. The business started in Berlin, Wisconsin under the name of Hardware Dealer Fire Insurance Company. In 1912, with P.J. Jacobs as secretary, the company headquarters moved to Stevens Point.

Just one year earlier, in 1911, Wisconsin passed the United First Workers Compensation Law. Employers were required to carry workers' compensation insurance as financial protection for employees injured on the job. Wisconsin Hardware Association took steps to form an associate casualty company affiliated with the fire insurance company. The new company, called Wisconsin Hardware Limited Liability Insurance Company, was granted license in 1914. The two companies began to expand coverages and lines. The casualty company adopted the name

Hardware Mutual Casualty Company, and first steps were taken to write insurance throughout the United States.

In 1922, a three-story office building in downtown Stevens Point was completed. Fire and casualty sales were combined in 1941. In 1956, an automated data processing system streamlined the renewal of automobile policies. Since that time, there has been steady progress in data processing , to provide better service to policy holders.

Another corporation of the company was formed in 1958, Sentry Life Insurance. This stock life insurance company was owned by Sentry's casualty and fire insurance companies.

In 1962, the companies chose the name Sentry, for a corporate image. The idea for the name came from the statue of Captain John Parker, located in Lexington, Massachusetts, which represented a minuteman in the American Revolutionary War. By 1963, the name Hardware Mutuals was dropped and replaced by the present Sentry Insurance name.

The commitment of Sentry to overseas operations began in 1967 when Sentry joined a group in London, England. Sentry also acquired operations in Australia and Asia. At this writing, Sentry's international group includes more than 45 affiliated companies operating in the United Kingdom, Australia, Bermuda, Hong Kong, Singapore, France, the Netherlands and other countries. A new building completed in 1977, constitutes Sentry World Headquarters.

Located in Portage County are more than 100 agents representing insurance companies with headquarters outside the county. Most of the agencies are located in the Stevens Point area, though in each village there are some. Insurance agents are often associated with banks or financial services as well.

EXPLOITATION OF NATURAL RESOURCES

The natural resources of Portage County are water, minerals and timber.

Water in Portage County can be examined in six areas: as a waterway for people and for floating logs; for consumption by humans and animals; as a source of energy for sawmills, gristmills and hydroelectric plants; for industrial use, especially in paper mills and for irrigation and for recreation.

The Wisconsin River has been used for transportation ever since the time of the Indians. Pioneers used the river at first for travel, then as a means of transporting supplies and logs. With the building of roads and railroads, less emphasis was placed upon the river.

Groundwater supplies water for human and animal consumption.

At first streams, especially the Wisconsin River, Mill Creek, the Plover River, the Flume River and the Tomorrow River, were used for furnishing energy for mills. But as water flow varied according to the seasons and the amount of precipitation, dams were built to form lakes (or flowages) and thus assure a reliable source of water.

In the manufacture of paper, water is considered a necessity.

Irrigation has been made possible through pumps which draw water from the ground.

Streams and lakes provide recreation in the form of fishing, boating and swimming.

Stone, gravel, sand and clay are the principal mineral resources of Portage County. Stone comes from two origins: erratics, brought by glaciers, and bedrock. Erratics in the eastern part of the county consist primarily of granite. In the western part, erratics are mostly granite mixed with gneiss and schists.

In Portage County, stones from a nearby area were used for building roads, foundations, dams and houses. Erratics have often been used for fences, especially in the eastern part of the county.

The earliest known quarry in the county is Connor's quarry located on the west bank of the Wisconsin River. Although the quarry has been abandoned for many years, its stone has been used for the early buildings of the city and for foundations and basements of houses of the nearby countryside.

A quarry near Ellis is noted for its buff-colored sandstone, marked by ripples in many layers. The sandstone was used primarily in early buildings, but today stone is taken from the quarry merely to match materials in existing buildings. Another sandstone quarry is Mosquito Mound, once used for building material before the turn of the century, but now abandoned. Some quarrying has been done in the Lone Rock area, where the stone was used locally.

The glacial outwash till and outwash plains have been formed from sand, gravel with pebbles and some boulders. Many pits have been opened throughout the county to furnish gravel and sand for roads and other construction. Some pits have developed into productive businesses. Sand and gravel are now an important resource, not only for Portage County, but for export outside the region. Since 1958, sand and gravel have been the most valuable mineral commodity in Wisconsin. Portage County is among 12 counties producing more than one million tons a year.

With development of the industry came a need for regulation. In 1967 the county established a planning and zoning regulation, which required a license for the exploitation of pits and specified that abandoned pits be reclaimed for other uses. The regulation however, does not apply to pits existing before 1967.

Two main active pits are those in the towns of Stockton and Buena Vista, which gather material out of the outwash plain of the Wisconsin glaciation. All pits operate in the same manner. The topsoil is removed and piled away. Prior to the law, the topsoil was sold as dirt or fill, but now the regulation requires that it be kept for reclamation use. A pit is dug. What is oversize is crushed. Sand and gravel are used for building, paving, founding metal and as railroad ballast. A large quantity of sand and gravel is shipped out of the county, especially to be used in cement plants in Marshfield or in the Fox River Valley.

Clay was used in brick-making almost as soon as European immigrants settled in the region. Though a brickyard was mentioned in area history as early as 1856, it was probably selling imported bricks.

A farmer who came from Germany and settled in the town of Hull in 1867, made bricks from clay on his farm. Production on a large scale started in 1880, and the first house built with local bricks was a mansion in Stevens Point. It was built for a German who operated the brewery there. It still stands today on Main Street and is locally referred to as "The Castle". This company, which produced about five million bricks a year, was one of the main suppliers for Stevens Point buildings.

However, there were other companies headed by Germans and Poles who had learned the art of brick-making in their native countries. Clay for the bricks was taken from creek beds, north and northeast of the city. The clay contained chunks of partially decomposed granite with varied cooling levels. Since the clay was not uniform, shrinking caused bricks to crack. Color also varied in shades of red. Finally, in 1923 the last company ceased production. Today the main site which furnished clay to the city is still visible; its deepest spots are filled with water. Another reason brick-making disappeared on the economic scene was that wood, a necessity in the process, became scarce.

Another mineral of importance in Portage County, was graphite. In 1893 a woman found a mineral in her cellar which she used to blacken her stove. It was analyzed as graphite. Several companies followed one another in mining graphite for the production of paste and paint. By 1900 several tons of finished paint and paste were obtained daily. But after a cyclone damaged a mill and a fire burned the rebuilt mill, operations were discontinued, and in 1921, the last graphite company was dissolved.

Portage County boasted a variety of trees, but it was the large stand of white pine that laid the foundation for the lumbering industry. Pines were so numerous, in fact, that the region was named "The Pinery". The lumber industry thrived from 1840 to 1880, when the woods of white pines were depleted. However, remaining trees were suitable for the paper industry. Today, most timber which is needed for paper mills and other industries is imported.

AGRICULTURE

Presently in the county, there are two barns of a special type: a round barn on Highway J near Custer, built in 1910 by a German immigrant and an octagonal barn, built about the same time and located in the town of Amherst. The round barn is characteristic of those built in the northern plains of Germany where there are strong winds; the round contour offers more resistance to air currents. The round structure also offers more space than one with four sides. In 1916, the College of Agriculture at the University of Wisconsin-Madison discouraged the construction of round or octagonal barns since they concluded the disadvantages of such structures outweighed the advantages.

Most barns and many sheds in the county are painted red. This is a custom brought from Europe where red paint, a by-product of iron ore, was cheaper than any other paint.

Improvements were costly but to be competitive, some farmers had to have the advantage of larger barns and stables. Some farmers expanded their farms and herds by buying out small farms whose owners would not, or could not afford to keep up with the trend.

Portage County followed the trend in Wisconsin from general farming to dairying but, because of the area's limited soils for growing hay, corn and oats, dairying was not as important as in other counties. In the county too, were many swamps, marshes, lakes and wooded areas. In 1890 the 2,626 farms that existed occupied 307,801 acres (124,562 ha). As immigration continued, more land that might have been used for dairying, was taken for farming.

Before 1902, approximately 30 percent of the southwestern part of the county (flat land that had been covered by glacial Lake Wisconsin) was still undeveloped. It was mainly wetland covered by marsh vegetation. From 1902 to 1904, land developers dug a series of east-west ditches at approximately one mile intervals, connecting with Buena Vista Creek, Duck Creek, Ten Mile Creek (all of which flow into the Wisconsin River). Land was drained and converted into farmland. In the north, more farms were created by clearing land left by lumbermen. Since much of the land was not suitable for hay, corn or oats, many farmers continued to raise potatoes as a cash crop, in addition to dairying.

In 1910, there were 3,229 farms covering 411,247 acres (166,425 ha): 49,960 acres (20,218 ha) were in hay; 15,834 (6,407 ha) in corn; 37,838 (15,125 ha) in oats; 1,184 (479 ha) in barley; 14,566 (5,895 ha) in rye; 397

(160 ha) in wheat; and 30,637 (12,398 ha) in potatoes. There were 51,336 heads of cattle, 13,264 swine, 5,075 sheep and 13,264 horses. In 1909, 273,505 pounds (124,062 kg) of butter had been churned on farms, and 2,127,143 pounds (964,870 kg) made in creameries and the three cheese factories in the county.

In 1909, the Bradley Polytechnic School in Peoria, Illinois, became interested in the Buena Vista Marsh as a vegetable farming area. The school had already established holdings in Illinois and the soil of Buena Vista Marsh looked promising for the same purpose. A drainage commission was formed, drainage ditches were laid out and brushing of the swamp was done by crews of White Russian laborers. After clearing the land, Bradley School set up a series of farms or ranches. Farmers from Iowa and Illinois became interested and set up a farming community, first named Pine Island, then Coddington after the first supervisor of the drainage work. An experimental research farm was established in 1916, but the soil was found to be unsuitable for growing vegetables, and the experimental station was closed around 1950.

The Depression of the 1930s kept people on the farm even after they reached an age at which they would have liked to retire. Land values had declined so that they could not sell their land and they were still able to obtain the major part of their living from the farm itself. World War II changed that picture. Demand for agricultural products raised both product prices and land values. Even though farming became attractive from these standpoints, many farmers preferred to sell their property.

From the first days of farming in Portage County, potatoes were raised on almost all farms because they were a staple food that could be cultivated in the climate of the region. In 1899, Wisconsin ranked third in the U.S. for the production of potatoes, just behind New York (the leading state since the agricultural census in 1850). Portage County ranked as the second county in the nation for acreage planted in potatoes. The 1850 census listed the name (and acreage planted in potatoes) of the 13 counties that had more than 15,000 acres (6,000 ha) planted in potatoes. Among them were Waushara and Waupaca counties.

Counties	Acreage	Production	Bushels/Acre
Aroostock, ME	41,953	6,466,189	154.13
Portage, WI	29,099	1,978,344	67.99
Steuben, NY	26,468	2,702,304	102.10
Wausahara, WI	23,685	1,905,737	80.46

Weld, CO	23,195	2,878,344	124.09

1 acre= 0.404 hectares
1 bushel= 0.352 hectoliters

The yield in Portage County was inferior to other counties and therefore, it was ranked fourth. The low yield per acre in Portage County (due to the topography and soil in some parts of the area) was balanced by other counties (Sheboygan-136 bushels; Dane-130 bushels; Milwaukee-127 bushels), so that the average yield for Wisconsin (95.90) always equalled the yield of New York (96.2).

In 1910, acreage, production and yield per acre had increased. Also the West entered the competition.

Counties	Acreage	Production	Bushels/Acre
Portage, WI	30,637	2,508,521	81.87
Aroostock, ME	75,738	17,514,491	231.25
Steuben, NY	30,524	3,279,953	107.45
Idaho(whole state)	28,341	4,710,262	166.20

1 acre= 0.404 hectares
1 bushel= 0.352 hectoliters

The following table shows the acreage planted in potatoes in Portage County from the federal census of 1962 and the Wisconsin censuses. Until 1949, production and yield were measured in bushels; after 1949, it was measured in hundred weights (cwt). It was decided that one bushel of potatoes was equal to a bag of 60 pounds or 0.6 hundredweight. Until 1977, only the acreage of harvested fields was published; later both planted and harvested acres were listed. The difference between planted and harvested acres is not large; the smallest was in 1983 (70 acres), the largest in 1978 (1,600 acres). The difference can be explained by heavy rains and floods and early frost, which prevented some fields from being harvested. For comparison, the table indicates very high and very low yields in other counties in Wisconsin.

1899	1,978,344	67.9 bushels
1909	2,508,521	81.9 bushels
1919	2,349,530	84.8 bushels
1929	1,941,830	86.3 bushels

1939	2,220,887	86.7 bushels
1949	1,382,286	95.0 bushels
1959	2,369,501	296.1 bushels (177 hundredweight)
1964	2,795,000	220 hundredweight (Trempealeau 90 cwt)
1969	4,452,000	280 " (Dunn, Monroe, Pierce 130 cwt)
1974	4,942,000	305 " (Adams, 354 cwt)
1979	6,339,900	336 " (Adams-378; Waushara-349 cwt)
1984	7,077,600	332 " (Adams-412; Waushara-417 cwt)
1986	7,000,000	348 " (Adams-446; Barron-392; Wood-407)

1 hundredweight (ctw)= 45.36 kilograms

Since censuses are sometimes revised, federal and state censuses do not always have the same figures. Also, because of varying weather conditions, yields vary. Through improvements in farming methods; mechanization; use of fertilizers, pesticides and products against disease; and removal of weeds--there was the first large increase in yield in 1899 (67.9 bushels; 113 cwt) to 1909's yield (81.9 bushels; 136 cwt). From that time until 1939, the yield remained in the 80's annually.

Although Portage County has an annual high rainfall suitable for agriculture, with the heat of summer the sandy soil does not retain moisture and potato growth is affected. In the 1930s, some farmers in the towns of New Hope and Alban irrigated their fields. Other farmers followed the example and in 1944, for the first time, average yield passed the 90 bushel mark (150 cwt). Acreage planted in potatoes was 8,962; five years later, acreage increased to 14,544 and the yield to 154.2 (257 cwt). It continued to increase to over the 300 hundredweight (cwt) mark in 1971, with a record 385 cwt in 1985. In the last 20 years, Portage County has shown the highest yield for some time, and production has remained the highest of all counties in Wisconsin. In countrywide ranking, Aroostock, Maine has remained first with 22,533,903 cwt and Portage County has dropped to 14th place, with 6,252,138 cwt. Adams County is 33rd, with 3,038,251 cwt. Production in Portage County is more than one-fourth of Wisconsin production, which has become the fourth in the country after Idaho, Washington State and Maine, and is followed by Oregon, California and North Dakota.

Irrigation fed by a good supply of groundwater is the main reason for high yields. Several types of irrigation are used. One consists of small sprinklers attached to aluminum pipes and moved by hand through the fields; the pipes can be carried from one field to another. In a second type, only the sprinklers are moved. Both systems require many pipes and are used for small fields or fields of irregular shapes. A third type is moved mechanically with the pipes serving as axles to a set of wheels; an engine moves the pipes and wheels. The most modern system is the pivot or circulatory system. It consists of a self-propelled pivot located in the center of a field next to a well. It turns with the pipes on wheels. It is fully automatic and fast, able to irrigate up to 140 acres (57 ha) in 20 hours. This system is the backbone of cash crops, potatoes and vegetables, but its use is limited to large square fields (such as those in outwash plains). Two areas of concentration of this system are: Bancroft/Almond and Plover, called the Golden Sands. However, the self-propelled pivot system can also be found on some farms in the towns of Amherst, New Hope and Alban.

Until the '70s, potatoes had to be sold in bulk. However, during that decade a processing plant did open, but closed because of lack of space for expansion and difficulty in disposing of residues. In 1974, the American Potato Company, with headquarters in San Francisco, opened a plant in Plover. The company produces potato granules, frozen French fries and dehydrated hashbrowns. In 1979, Ore-Ida Foods, with headquarters in Boise, Idaho, constructed a plant in Plover. The company produces various kinds of frozen potatoes. They had plants in Oregon, Idaho and Michigan, but in 1987, the plant in Michigan was closed.

To obtain high yields, farmers must rotate crops. Potatoes may be planted every other year, or every three years or even every four years, as secondary crops. The fields may be planted with corn, which is used to replenish the humus in alfalfa, which will supply nitrogen to the soil as a legume; or in rye, to be used as forage for cattle. A number of farmers grow potatoes to feed livestock, either for production of beef cattle or dairy cattle.

In early days, no distinction was made between beef and dairy cattle, but eventually purebred animals were introduced. At present, the most common breed is Holstein, but there are also herds of Jerseys, Guernseys, Ayrshires and Brown Swiss for milk, and Hereford and Angus for beef.

On January 1, 1986, there were 44,000 heads of cattle and calves: of these, 17,900 were milk cows; 2,500 beef cows and the remainder were calves, heifers, steers and bulls. The milk cows gave an average of

13,600 pounds (6,168 kg) of milk a year, slightly higher than the state average of 13,383 pounds (6,070 kg).

According to the 1985 Wisconsin census, cash receipts from farm marketing amounted to $97,412,000, distributed as follows:

Dairy products	$28,989,000
Meat animals	13,768,000
Poultry and eggs	291,000
Miscellaneous livestock products	84,000
Field crops	7,651,000
All vegetables	43,451,000
Other crops	3,178,000

Portage County held first place in the state in the category of "all vegetables", represented mainly by potatoes. In 1985, the acreage planted in potatoes (20,450 acres--58,275 ha) was less than one-tenth of farmland (287,000 acres--116,185 ha). The Wisconsin census does not give per county acreage of all crops. In Portage County, as in the entire state, corn and hay occupy the largest areas: 1986 - 48,100 acres of corn for grain; 7,800 acres of corn for silage; 3,800 for sweet corn; and 47,400 for hay (almost half of which is planted in alfalfa). Other crops listed are as follows:

Oats	- 9,500 acres harvested, but 13,100 planted 9,900 planted and 9,100 harvested in 1985
Soybeans	- 1,100 acres harvested
Barley	- 600 acres harvested
Wheat	- 400 acres harvested
Snap beans	- 17,800 acres haves
Green peas	- 4,200 acres harvested

The acreage of other crops is only listed for the whole state. Crops found in the county are:

Mint for oil	- 9,800 acres
Cranberries	- 7,900 acres
Cucumbers	- 7,200 acres
Rye	- 6,000 acres
Strawberries	- 1,300 acres

Onions	- 1,200 acres

The federal census of 1982 gives a few more details, but it is not fully reliable as it is based on samples and there is a sizable discrepancy between the two censuses. For instance, in 1982, the federal census shows 13,072 acres planted in snap beans and 1,949 in green peas. The Wisconsin census lists 14,600 for snap beans and 3,300 for green peas. However, data from the 1982 federal census gives some information of the importance of a few crops in the county:

Rye	- 1,110 acres
Cucumbers	- 696 acres
Sorghum	- 22 acres
Apple trees	- 10 acres

Mint was introduced in the Bancroft region in the 1940's but, because it covered only a small area, Portage County was not indicated among the counties raising the crop. Cranberries, onions, strawberries and grapes were listed among Portage County crops, but production figures were withheld to avoid disclosing data of individual farms. There were only two cranberry growers and together their bogs covered somewhat more than 200 acres (81 ha).

According to the 1986 census, in addition to dairy and beef cows, other animals on farms were: 14,700 hogs and pigs, 2,500 sheep, 22,600 laying hens (producing 6,380,000 eggs) and chickens, turkeys, goats, rabbits and bees. The 1982 federal census gave these figures: 34 goats; 140 colonies of bees providing 17,240 pounds (7,820 kg) of honey for sale; 176 ducks; 28,840 chickens and 11,642 broilers (the number of turkeys was not disclosed, but there were 18 in the 1978 census). There were also 30 ducks and geese and 25 rabbits.

Snap beans and green peas were productive crops in the outwash plains because of the soil conditions and the availability of an excellent supply of groundwater for irrigation. Production increased with the introduction of the mechanical snap bean harvester in 1955. At one time farmers used migrant workers to harvest these crops. Today, processing is done at the cannery established by the Del Monte Corporation.

Corn and various grains are used to feed cattle. Corn, as ripened grain, is ground in a blend formulated from oats, beet pulp, molasses, salt, bone meal and other high protein feeds purchased from mills. Corn can also be fed as silage, which is the entire corn plant harvested while green, and stored in silos or pits. Most dairy farms have silos--generally one per farm, but often two, three or more. Alfalfa, timothy and clover are grown for hay and stored in barns for the months that cattle cannot graze on pastureland.

Maple syrup is listed as a product in the 1964 federal census; that year, 48 gallons (181 l) were sold. The 1964 census lists 8,144 Christmas trees sold by 21 farms. Subsequent censuses do not mention these products.

Other farms not mentioned in agriculture censuses are game, birds and fowl farms. Such establishments need a license from the Department of Natural Resources. At this writing there are approximately 30 licensed holders in the county--but only two, near Nelsonville, raise pheasants and have commercial game farms. These pheasants are available to gun clubs; other license holders have game farms as a hobby. There are a variety of animals raised as game: squirrels, rabbits, pheasants, partridge, grouse, quail and peacocks. A number of license holders have no active farm.

A family near Custer raises, as a hobby, a large variety of domestic and exotic animals, such as: buffalo, deer, fox, llama, bear, peafowl, ducks and emus from Australia. Surplus stock is sold to zoos or other game farms.

Federal censuses provide great detail about agricultural products and farms, but since the last published census was in 1982, some figures are outdated and many tables are not of general interest. It is, however, worthy of note that in Portage County the majority of farms are operated by their owners. In 1982, there were 1,119 farms, 683 were fully owned and operated by one man or one family. There were 375 part-owners, and only 61 tenants. Of the farm operators, 954 lived on the farm, 86 did, and 79 did not report their place of residence. Of these farmers, 737 indicated farming as their principal occupation: 525 stated they did not spend one day of work off the farm, and 620 reported to have been more than 10 years on the same farm. Only 25 operators were 25 years old or younger. The numbers increased as age increased, peaking at 263 operators between 55 and 64 years. The average age was 48.5 years.

According to the 1986 Wisconsin census, Portage County ranked first in the state in potatoes. 22,400 acres (9,065 ha) of potatoes planted, with a production of 8,633,000 cwt (93,916 million metric tons). The figures were more than double those of the next county (Langlade - 3,423,000 cwt), which was followed by Waushara and Adams counties. However, in Portage County, the yield per acre, 385, ranked only sixth after Adams

(486), Barron, Juneau, Waushara (450 each) and Wood (400). Portage County ranked first for snap beans (19,500 acres--7,891 ha.) harvested, producing 57,600 tons. The second was Waushara with 13,300 acres and 41,900 tons.

The 1982 federal census ranks Portage County as second for snap bean acreage (13,072), not far behind Marion, Oregon, which was first with 13,824 acreage. However, in the previous census, although Portage County had a higher snap bean acreage, it ranked fourth after Marion; Palm Beach, Florida and Waushara. In potato acreage, Portage ranked 14th.

In the 1986 census, Wisconsin ranked fourth for potato crop; Idaho was the leader, with Washington and Maine taking second and third respectively. Oregon placed fifth behind Wisconsin. Wisconsin ranked first in snap beans, green peas and sweet corn. The state was first in dairy production, first in the production of milk, butter, cheese and whey products, in corn for silage and in alfalfa hay.

INDUSTRY

Manufacturing

The first people to make their living from resources of the land were lumbermen. They were attracted by the large stand of pines, especially white pines.

Wood was one of the resources used by immigrants as building material for their homes, furniture and as fuel for heating their lodgings and cooking food. Immediately in demand by America's first settlers were big trees to furnish good boards and softwood. Maine was the first state to be logged, followed by other New England states, New York State and Pennsylvania. As the population increased and spread out, new sources of wood were sought. The rich reserves of the Midwest, particularly Wisconsin and Minnesota, became of interest.

Forests at that time were not inhabited since the Indians, few in number, lived in the open areas. The first lumbermen came by way of streams and lakes since these were the only avenues to connect forests with the outside world. At first logs were rafted down streams until they reached towns with mills where they were processed. Soon it became evident however, that by sawing logs closer to their place of origin, damage that occurred when logs were floating through rapids could be eliminated.

Soon after the Treaty of 1836, which opened land along the Wisconsin River to settlers, lumbermen came to the region that would become Portage County to establish sawmills and use water from the streams for power. Mills were located on the shores of the Wisconsin River. The first mill was established in 1837 on the west bank by Gilbert Conant and Daniel Campbell at Shaurette Rapids. A second mill was established by Azon Richardson on the east bank at the site of what is now Stevens Point. In 1839 two mills began operation on Mill Creek and on the Plover River. The latter was dammed, forming Jordan Pond. In the next decades other mills were built, especially in the vicinity of Stevens Point which became the center of the lumber industry. At least six mills were on the banks of the Wisconsin River. Other mills were: one near the mouth of the Plover River, three upstream of the Plover River two on the Tomorrow River and four on the Flume River. The last mill was built in Rosholt in 1885. These mills were the first industry in the county, producing boards and shingles.

Following the lumbermen, farmers came to the area and settled in open lands east and south of Plover and Stevens Point. Gristmills were established for grinding grain grown by farmers. Planing mills were also constructed. It is believed the first was in Stevens Point in 1874. Later, in 1879, there were seven such mills.

With the construction of railroads, the lumber industry developed. Since there was ample lumber, some men saw an immediate market for the fabrication of building materials, such as doors, windows, roofing material and furniture. When production exceeded what was needed in the area, export started. In order for industry to prosper along the Wisconsin River, the river itself had to be controlled. This was achieved by building a series of dams and spillways which allowed industry to harness power, control flooding and drain swampland.

By the end of the nineteenth century local forests were depleted of trees considered good quality for lumber, and the lumber industry declined. But, left behind were varieties of trees considered very good for making paper. Consequently, there was a shift from lumber to paper along the Wisconsin River. The first company, the Plover Paper Company, was formed in 1891 by George A. Whiting and his associates. The first mill built was taken over by George Whiting and, in 1912, the name was changed to Whiting Plover Paper Company. Although in recent years, the mill has changed ownership, it is still active. In the same year of 1891, another mill was started by the Wisconsin River Paper and Pulp Company. The first mill was known as the lower mill and the latter as the upper mill. The upper mill was eventually purchased by Consolidated Water Power and Paper Company of Wisconsin Rapids in 1945. It took the name of Wisconsin River Division of Consolidated. In the meantime, in 1917, Consolidated Company had established a new mill in Stevens Point with a dam at Shaurette Rapids. This became the Stevens Point Division of Consolidated.

At present, each of these three mills specializes in producing certain types of paper. The Stevens Point Division of Consolidated Papers is the nation's largest producer of high-tech coated specialty paper used for packaging foodstuffs, apparel, soap, beverages, detergents, cosmetics, pharmaceuticals and chemical products. The Wisconsin River Division is a leading producer of lightweight, coated printing paper for the nation's prestigious business and consumer publications. Its main market is in the Chicago area where there is a high concentration of magazine printing companies. Consolidated's paper is used in the following publications:

Cosmopolitan, Forbes, Fortune, Good Housekeeping, Harper's, Money, Newsweek, Omni, People, Playboy, Reader's Digest, Science Magazine, Sports Illustrated, The New Yorker, Time, TV Guide, U.S. News and World Report,as well as the catalogs of Montgomery Ward, J.C. Penney, Sears and others. The Neenah Paper Whiting Mill makes premium correspondence paper. It uses cotton fiber and sulfite wood pulp in proportions ranging from 100 percent cotton to 100 percent sulfite wood. The company produces bond and writing papers, texts, covers, ledgers and onionskin paper--all of different colors and finishes.

In 1893 Herman A. Vetter, a resident of Stevens Point, went to the Chicago World's Fair. While riding a streetcar there, Vetter met a man who owned property in Stevens Point. The man asked Vetter if he would run his Stevens Point mill. Vetter accepted. He bought some woodworking equipment, hired three men and was in business. At first Vetter Manufacturing made cabinets, stairways, blinds and doors, but today the company is solely a window unit manufacturer.

The turn of the century marked the beginning of furniture manufacturing in Portage County. In 1897 John J. Bukolt, a Polish immigrant, patented his invention of an automatic cradle for children, and he began to manufacture such cradles. In 1904 Bukolt founded a company--Automatic Cradle Manufacturing. In 1920 the name was changed to Lullabye. The company has changed ownership again and it is presently a concern that the manufacturer may leave Stevens Point for another location.

In 1899 the Coye Furniture Company was founded to manufacture furniture. It was taken over in 1917 by Charles A. Joerns and today was operating under the name of Joerns Health Care Company.

Another industry which started at this same period was that of making fishing tackle. Jack Frost, a Stevens Point resident, enjoyed fishing trout and made his own trout flies. He taught his daughter and in 1896, there began a small industry which later became Frost Fishing Tackle Company. In 1921 Oscar L. Weber, who had been manager of the company, resigned to start the Weber Lifelike Fly Company. In 1926 it merged with Frost's company. Other fly-tying companies began, among them the Worth Company. In the course of time, Weber began to manufacture other products than those associated with fishing. The company started using plastics, and it became Weber Plastics. In the 1940's approximately ten million flies were produced in Stevens Point. It could have been designated as the fly-tying capital of the world, but the introduction of the spinning rod caused the decline of the fly-tying industry. Weber closed his factory in 1985 and at this time, only Worth Company remains in operation.

Another local industry is printing. It had its start in 1898, when Worzalla Publishing Company was formed. It took over editing a weekly Polish newspaper, the Rolnik, and in 1908 began publishing the Gwiazda Polarna (The Northern Star). Worzalla published books and in 1932, added another activity with the organization of the National Bookbinding Company. Another printing company in operation is the Journal Printing Company.

The aforementioned industries were focused on the large markets, but from early days there have been craftsmen and small industries aimed at the local market. In Stevens Point in 1856, there were three blacksmith shops, six carpenter shops, one harness shop, two clothing shops, three boots and shoes shops, two millinery shops and one meat market. Employees numbered 30 carpenters, ten brick layers or stone masons, five painters, eight blacksmiths, six wagon and carriage makers, eight cabinetmakers, seven shoemakers and eight tailors. There were also shops in Almond and other places.

Because Germans, and later Poles, formed an important ethnic portion of the population, their custom of drinking beer influenced the establishment of a brewery. Frank Wahle and George Ruder were the first owners. In 1857 an expansion of their business was announced in the local newspaper. A second brewery was started in 1866 by Adam Kuhl, also a German. The business continued until 1925. Several other breweries existed in Stevens Point before prohibition but none lasted for more than a decade, and they made beer for the local market only. Only the brewery founded by Wahle and Ruder is in existence today. It has changed ownership several times. In 1902 it was organized under the name of Stevens Point Brewing Company. During prohibition, the company was reorganized and named Stevens Point Beverage Company, producing near-beer, a beverage that resembled beer. It had little success. The company managed to survive and resumed making beer as soon as prohibition was repealed. It was one of the 725 breweries that operated in 1934 in the country. It resisted attempts to be bought up by larger companies, and did not try to expand. Point Special Beer is recognized for its quality. Following recent trends of larger brewers to add light beer to their production, the Stevens Point Brewery in 1985, added Eagle beer, a light beer. The company believes it will continue to operate and that there is a place in the market for its products.

In other parts of the county, other industries were begun. When Rosholt developed as a lumbering center, logs were shipped to Schofield and Chicago. Later as Rosholt businessmen attempted to provide employment for seasonal labor, the Rosholt box factory was started. At first the factory manufactured cheese boxes but shortly switched to manufacturing wooden pallets used as platforms for fork lift trucks. In 1954 the Doede Manufacturing Company, which also makes pallets, was organized.

When dairying began to develop, creameries began to operate all over the county. Most creameries produced butter; some produced cheese. But as transportation lines improved, larger plants outside the county became the outlets for milk.

Until 1950 almost all industries were organized and operated by local individuals or families. Stevens Point, in a move to attract new industries to the area, formed the Industrial Development Corporation. There were 250 stockholders. Funds were used to purchase vacant land in the area of Rice, Michigan and Helen streets south of the Soo Line railroad, which provided a spur. The area was zoned "light industrial." Freda Sausage was the first resident company, followed by Al-Chroma, a manufacturer of paints for metal and aluminum structures. Later a sister of the Industrial Development Corporation, namely the Industrial Development Foundation, was organized to assist other industries settling there. The first was Steel King which produced industrial jacks and steel buildings for gas stations. It was followed by Duralum, a company that made siding and roofing materials. Okray, a potato processor, also began there but closed after a few years, largely due to problems of eliminating waste through the sewage system.

In 1966 Del Monte, headquartered in California and one of the largest fruit and vegetable canning companies in the country, opened a canning plant in the town of Plover next to the Green Bay railroad track. At present, canning is limited to green beans of different varieties.

In 1965 Joerns Furniture was anticipating a move to larger quarters, or possibly a move out of the area. To keep the industry, the city of Stevens Point, in 1967-68, bought a parcel of land containing 45 acres (18ha) located on the east side of the city south of Highway 10. The parcel was paid for from the city's general fund and marked the city's first entrance into purchasing land for industrial development. Through efforts of the

State Economic Development, Herrschners, Inc. was persuaded to locate in the industrial park. The company was established in 1899 in the Chicago area. The reason for coming to Stevens Point was partially for more space since the Chicago building was old and there was no possibility of expansion. Another reason for Herrschners' relocation was that there was an abundant workforce here willing to work full time or seasonally for lower wages because of the area's lower cost of living. The company makes material for needlework, and its specialty is needlecraft kits distributed to various parts of the United States and Canada. When Herrschners came to Stevens Point in 1970, it was the first company operating in the industrial park. The workforce fluctuates since primary sales are via mail-order. The yearly catalog comes out in early December, orders increase until May and then, decrease. Consequently the number of employees may reach a peak of 230 in winter months, and decrease to approximately 150 in the summer. The company also has a small factory outlet on the premises.

Sirco, a manufacturer of small sewing machine tables, cabinets and office furniture, came to the industrial park from Montana in 1971. The company wanted to be closer to the major market for their products. They have about 50 employees.

Joerns Furniture Company bought land in the industrial park, plus a lot next to it, and obtained a spur track from the Wisconsin Central Railroad (formerly the Soo Line). In 1970 the company built a large plant. It now specializes in making furniture for hospitals and for this reason, the name was changed to Joerns Sunrise Medical Inc. It has contracts with the government.

At first the main problem of the industrial park was one of access. The main access from Highway 10 though Country Club Drive and Hoover Road has largely been solved by building a bridge over the Plover River which joins Industrial Park Road to Patch Street, leading to Michigan Avenue and Business Highway 51.

In addition to manufacturers, the industrial park has distributors like Midstate Distributing Company for beer and soft drinks, and NAPA for automotive parts. Two companies that began in the city have moved to the industrial park: the Journal Printing Company was previously next to the Stevens Point Journal in a building that had to make room for the new Highway 10, and Colorcraft, which processes and prints photographic color films and employs about 175 people.

Increase in the production and quality of potatoes induced large companies to establish plants for processing these tubers. The first was the American Potato Company, of the Potato Division of Basic American Foods, a company headquartered in San Francisco. The potato division consists of four establishments: the main one in Blackfoot, Idaho. The others are in Rexburg, Idaho, Moses Lake in Washington and the one in Plover which opened in 1973. Sunspiced, a subsidiary of Basic American Foods, buys the potatoes and grades them. Grade A is sold fresh while grade B is used to prepare dehydrated potatoes in the form of granules and hashbrowns. Among clients of American Foods are food service companies and restaurant chains.

In 1979 Ore-Ida (Oregon-Idaho) Foods, Inc., which had started in Boise, Idaho, opened a new plant in the Town of Plover. It employs 750 workers and specializes in frozen food, especially French fries and hashbrown potatoes. The corporation has plants in Ontario, Oregon, in addition to the one in Plover. A Michigan plant recently closed.

A newer company, Celestial Farms, Inc. started processing potatoes in 1983. It is located in the Village of Plover next to the Green Bay and Western Railroad. The company employees 250 persons year-round. Celestial Farms produces only frozen potatoes, either whole or stuffed, potato shells and strips. Many of the potatoes are used by other food companies in conjunction with vegetables or meat.

In addition to the industries discussed, there have been other industries established in Portage County, especially in the Stevens Point/Plover area. As of this writing, the Chamber of Commerce lists 48 manufacturing companies.

Grouped according to number of employees, are:

Large establishments: Six companies with more than 250 employees, and seven companies with from 101 to 250 employees.

Medium-size establishments: Seven companies from 51 to 100 employees, and six companies with from 26 to 50 employees.

Small establishments: Eight companies with from 11 to 25 employees, and 14 companies with from three to ten employees.

Types of industries were listed in descending order according to the number of each type of industry:

Printing, publishing	12
Food products, beverages	10
Equipment (industrial, agricultural)	9
Furniture	3

Paper	3
Software	2
Asphalt	2
Bleaching compounds, concrete blocks, fishing tackle, needlecraft, paint and film processing*	

*one company of each

Of 13 large establishments, as many as seven were started because of the quality and abundance of local resources. Access to timber was the reason Joerns Healthcare began producing furniture here. Timber and water were the base of paper fabrication and the reason paper mills were founded in Stevens Point, the Wisconsin divisions of Consolidated Paper with headquarters in Wisconsin Rapids, and the Neenah Paper Company. Agricultural products, especially potatoes and green beans, explain the origins of American Potato Company, Ore-Ida Foods and Celestial Farms.

Printing companies were a natural outgrowth of the paper industry. Worzalla Publishing is a large establishment, Journal Printing is among the medium-sized companies and there are ten additional printing establishments. Area lakes and streams probably stimulated the founding of fishing tackle companies, not only in Stevens Point but also in nearby Amherst. Three large establishments, Herrschner, Steel King and Woodward Governor, settled in the area because of the abundance and quality of the labor force and the location, availability and low cost of land.

Steel-King began as a small company, but eventually expanded and diversified its products. Presently there is a main factory in Stevens Point with 200 employees and two factories in New London with 80 employees. The company produces industrial containers, storage racks and steel guards for work areas.

Woodward Governor's corporate office is in Rockford, Illinois. In 1870 Amos Woodward received a patent for a governor that controlled the flow of water onto a waterwheel. Subsequently Woodward founded a company which specialized in the design and manufacture of controls for waterwheels and water turbinds. In 1933, the company began to produce governors for diesel engines and aircraft propellers. Beginning with a membership of 12 employees and some 16 production machines, the company grew to more than 330 employees with 90 production machines.

The Worth Company is the only remaining company of several fishing tackle establishments that were founded in Portage County. The company survived because it adapted to modern ways of fishing by making a variety of lures and various types and colors of spinners, spoons and jigs. Another reason for the company's success is its production of accessories: a fish scaler, depth finders and anchor controls. The Worth Company has 55 employees in Stevens Point and representatives covering the international market and Canada.

In the early 1960s, one man began developing and printing photographic film for amateurs in a company he named Mid-State Photo Laboratory. In 1968 the company had grown, employing from 10 to 25 employees. In 1970 the name was changed to Colorcraft, printing color film as a specialty. The company grew rapidly to 177 employees, becoming one of the largest establishents of its kind in Wisconsin. Today it goes under the name of Qualex, Inc.

Among the middle-size and small companies, many produce for the local or regional market, but some make unique products. Foremost, with 50 employees in Plover, produces food additives from whey. Al-Chroma (Al for aluminum and Chroma for color) was founded in 1956 when one man developed a method of eliminating the dark brown pigment from asphalt. This resulted in more attractive uses for asphalt-based coatings. With an aluminum fiber additive, the feature of strength was realized and the product became a natural for roofing material. Al-Chroma began to market its products nationwide. In the early 1970s, the boom of tennis as a sport opened up a large new market for Al-Chroma's multi-colored coatings. When that market was satisfied, the company focused on roofing and the tennis court maintenance business.

Although it employs only 13 people, So-White Chemical Company in Plover, which opened in 1946, is Wisconsin's largest manufacturer of household bleach.

The Rosholt Stained Glass Company began in 1973 in Fond du Lac. It was moved to Rosholt when the building of a new Lutheran Church there left the old church abandoned and with ample space for the industry. The company consists of only one man who is owner, manager and worker.

A most unusual company, perhaps unique in the nation and the world, is Magic Products, Inc. (previously named Magic Worm Bedding Company). The bedding is a mixture of sphagnum peat moss and other ingredients which fishermen use to keep worms healthy. The company makes worm food. The promoter opened a laboratory where he studied worms and their habits, and developed a special bedding that became the base of the company. The company attracted the media, and the man appeared on television to describe this unusual industry. The company now employs eight.

The U.S. Census of manufacturers shows the evolution of the industry. The most recent census dates back to 1982, and is already obsolete since many changes have taken place. Indeed censuses cannot be fully comparable since definitions and classifications of manufacturers has changed.

The 1860 census lists the following industries:

	# of companies	male/female		total
flour/meal	1	2		2
leather	2	2		2
sawed lumber	6	90	6	96
printing	2	4		4

The 1880 census lists:

	# of companies	male/female		total
gristmill products	8	25		25
foundry/machine shop	3	25		25
planed lumber	4	54		54
sawed lumber	31	337	11	368

In the 1900 census there were 107 establishments with a total of 1,123 employees: 163 were proprietors and firm members; 64 salaried officials, clerks, etc. and 1,123 wage earners of which 940 were male, 163 were female and 20 were children under 17 years.

The 1905 census indicates only 37 business establishments in Stevens Point. The explanation given is that a number of establishments have been omitted since they are no longer considered manufacturing. There are 60 types such as: blacksmith shops, confectioners, grist mills, dairies, quarries, drugstores, locksmiths, monument work; in general most repair shops or shops producing exclusively for local consumption. Because of this, only 37 establishments were listed for Stevens Point; in 1900, there had been 107.

The 1939 census gives 57 manufacturing establishments in Portage County with 1,260 wage earners; 32 companies with 812 wage earners in Stevens Point. In 1947 in Portage County, there were 52 establishments with 2,361 employees, 2,056 of them production workers; and in Stevens Point 31 establishments with 1,681 employees of which 1,458 were production workers. In 1958 in the county there were 63 establishments, only 15 had 20 employees or more; in Stevens Point were 28 establishments, ten with 20 employees or more.

The 1982 census lists 63 establishments in the county; 39 in Stevens Point and 12 in Plover. In the county 27 employed more than 20 persons; 19 in Stevens Point and 6 in Plover. There were a total of 4,000 persons employed in the industry. The payroll amounted to $72.2 million. Among them were 3,100 production workers with a payroll of $50.3 million.

From the time the first sawmill was built in Portage County, industry underwent a number of changes in the products manufactured, in the development of several branches and in the decline of other industries.

From the 1968 list of industries in Stevens Point prepared by the Chamber of Commerce, only 18 manufacturing companies still existed in 1988. Thirteen have the same name: Acorn Equipment Corporation; Al-Chroma, Inc.; Consolidated Papers, Inc.; Stevens Point Division and Wisconsin River Division; Del Monte Corporation; Freda's Sausage Company; Gamber-Johnson, Inc.; Journal Printing Co.; So-White Chemical Co.; Stevens Point Brewery; Stevens Point Concrete; Worth Company and Worzalla Publishing Co. Five companies have new names: Colorcraft Corp., Inc., now Qualex, Inc. (Mid-State Photo Lab); Evenflow Juvenile Furniture (Lullabye Furniture Co.); Joerns Healthcare, Inc. (Joerns Furniture Co.); Neenah Paper Co. (Whiting -Plover Paper Co.) and Steel-King Industries (Mercury Mfg. Corp.).

Among factories listed in 1968 but not in 1987, is Haertel Monument Service which is still in existence but probably omitted from the guide as the U.S. Census does not consider this type of business as a manufacturer. This establishment was founded in 1901. When Stevens Point celebrated its centennial, Haertel received a certificate of recognition for having been in business in the community, and in the same family, for 50 years or more. Such certificates were given to Journal Printing Company, Joerns Brothers Furniture, Worzalla Publishing, Company Lullabye Furniture and other businesses that were not manufacturers.

Many industries started in Portage County; some expanded, some remained small, others disappeared because the business was not profitable or because the owners moved out of the area. Since 1968, two establishments that at one time had been important, have since closed: Weber Plastics and Bake-Rite Company, the latter having employed up to 300 people.

Although Stevens Point/Plover is the industrial center of Portage County, there are a number of industries that have developed in other parts of the county--some having started just recently.

In Amherst Junction (in addition to Magic Products "For Happy Worms" as is written on the building), Pavelski Entreprises dates its beginning to the mid-50s. The company calls itself "crop care specialists," and manufacturers fertilizers adapted to the soils and crops of individual farms in five counties. In the peak of their season, the company employs 85 people. In Amherst, there are two printing shops and a chemical factory, Secodyne, which makes products used in sewage treatment plants. The latter was founded in 1984, and is a branch of a company headquartered in Detroit, Michigan. In Nelsonville is Rodmar Manufacturing and Research, Inc., founded in 1981, which makes custom plastic moldings that are used in a variety of industrial products. Forty employees work in two shifts.

Portage County strives to attract new industries. In 1987 an inventory of industrial sites was made and 86 available sites covering 3,769.3 acres (1,525 ha) were identified: 16 in the Village of Plover with an acreage of 837.01 (339 ha); 11 in the Town of Plover with 584.23 acres (236 ha); 13 in the Town of Grant with 243.04 acres (983 ha); 12 in Stevens Point with 13,767 acres (5,571 ha); ten in the Town of Hull with 761.11 acres (308 ha); six in the Town of Stockton; five in the Town of Amherst; two each in the towns of Buena Vista and Sharon; one each in the towns of Carson, Lanark and Pine Grove. The Village of Plover now has an industrial park covering 103 acres (42 ha) on Highway B.

TEN LARGEST MANUFACTURING EMPLOYERS 12-21-86

Company Name Product or Business Employment

Ore-Ida Foods, Inc.	Frozen Vegetables	762
Consolidated Papers, Inc.	Paper Products	757
Woodward Governor	Governor Manufacturer	375
Joerns Healthcare, Inc.	Hospital, etc. Furniture	304
Worzalla Publishing Co.	Commercial Printing	260
Qualex, Inc.	Photo Finishing Lab	240
Kimberly-Clark	Paper Products	216
Herrschners, Inc.	Mail Order House	153
DelMonte Corp.	Vegetable Processing	147
Steel-King Industries	Industrial Shelving	124

Diversifiction of industries explains why the employment rate is relatively low in Portage County. When one company lays off workers or closes, employees--especially the young--have little difficulty finding another job.

NON-MANUFACTURING

Construction was one of the first activities practiced among newcomers to Wisconsin. It was the trade of carpenters, with laborers as helpers and occasionally other specialized workers. Most likely involved also, were home owners and their families.

The 1850 census lists 10 carpenters and one painter in Stevens Point and 18 carpenters in the Town of Plover. No other craftsmen seem to have been involved in construction.

As time went on, other craftsmen in addition to carpenters became involved in construction. Today the industry includes many types of work and various types of businesses. Building is the task of contractors who use specialists, often drawing them from other counties or even other states.

At present the construction industry is divided into three categories:
1. Building contractors in charge of residential buildings, industrial buildings, warehouses, public buildings, schools, churches, institutional buildings, office buildings, garages and farm buildings.
2. Heavy construction contractors primarily engaged in building highways, streets, bridges, sewers, canals, power lines and irrigation projects.
3. Special trade contractors whose activities include painting, electrical equipment installation, plumbing, heating, air conditioning, applying siding and roofing, excavating, well-drilling and installing irrigation systems. They may work as subcontractors to the general contractor, or they may work directly with the owner. For the most part, special contractors work at the building site, although they may have shops where they perform jobs incidental to the job site.

Although the construction industry does not employ many people directly, it is vital to the economy since it uses the services and products of other industries. According to the County Business Patterns published by the U.S. Bureau of the Census, in Portage County in 1985 there were 110 establishments dealing with contract construction: 76 employed from up to four persons; 24 employed from five to ten; nine employed from ten to 19 and the last construction group employed more than 19 employees. There were 71 special contractors of which 18 were plumbing, heating and air conditioning; 16 masonry, stoneware and plastering and the remainder involved a number of other skills.

Central Wisconsin is home to the Golden Sands Home Builders Association with 180 members in Stevens Point and Plover, but also with members in Wood, Marathon, Waupaca and Waushara counties. The list of members published in 1987 indicates there were members other than those connected with the construction industry: representatives of banks, insurance, real estate, abstract companies and building inspectors. For Portage County there are 38 building contractors: 11 representatives of the cement and masonry industry; seven in the electrical business; six dealing in lumber and like products; four each in excavating, painting and utilities; three each in floor covering, insulation and kitchen cabinets and one each in roofing, well-drilling and overhead doors. Some do not work exclusively for construction, but may have repair shops or retail stores. Most are in the Stevens Point area, although there are some in all villages (incorporated or not) and several live in the country where their home is also their place of business.

The trucking industry in Portage County is represented by several types: common carriers, which handle boxed or packaged freight; bulk carriers, carrying either liquid (fuel, milk, chemicals) or dry carriers (sand, gravel, cattle feed, truck crops, paper, lumber). The industry is regulated so that companies operate in zones: some are interstate, their zone comprising several states or even the whole country; others are intrastate, operating inside the state while others are local and do not extend outside the city.

Regulated companies are firms which have a gross income of one million dollars or more. They have assigned areas designated by the Interstate Commerce Commission and by the Wisconsin Public Service Commission. The areas of service are to various cities along prescribed routes. If an article of freight is destined for a city that is not on the company's route, it will be turned over to a firm that does cover that city. New areas of service may be acquired through purchase of territory from another company. Another way is, in recognizing a need for service in an area, receiving rights through a hearing of the Wisconsin Public Service Commission. However this is possible only when the area borders a company's current routes.

Each company has several terminals in its service area. Freight is picked up by a local truck, taken to the nearest terminal and reloaded to another truck. Then it is taken to the terminal nearest the destination point and reloaded to a local delivery truck. If the destination point is not along the routes specified for a certain company, it will be transferred to a second company's terminal. Trucking companies work well with each other to provide rapid service.

Moodie in Plover and H.O. Wolding in Amherst have their headquarters in Portage County. Con Way Central Express, Roadway Express and Yellow Freight System have terminals in the Stevens Point area. They are common carriers. All are interstate companies. In addition, there are several heavy hauling trucking companies and some local trucking companies in the Stevens Point area.

The field study for this section was made in 1972. Since that time, deregulation has taken place but it has had little effect on the trucking industry. Deregulaltion, however, has made it easier for trucking companies to acquire other companies or new routes.

TRADE

John Baptiste DuBay was the first trader to come to the territory of what is now Portage County. His father, Jean Baptiste DuBay, a French Canadian, married an Indian woman who gave birth to his son, John. There are discrepancies as to the year and place of John Baptiste DuBay's birth. Some sources say he was born in Green Bay in 1808; another source indicates Detroit, 1802, and still another source, 1810.

John Baptiste DuBay married Madeline, "The Princess," who was said to be the daughter of the Indian, Chief Oshkosh. DuBay opened a trading post on the Wisconsin River, said to be in 1830. However, the first documentation is of a contract dated September 7, 1840, by which the American Fur Company of the Astor Syndicate consigned "goods, and other ware and merchandise" in exchange for "money, furs, and other returns." As immigrants became attracted to the region for its lumber, DuBay (having acquired some wealth) became a lumberman himself. He retained his position as head of the trading post. The 1850 census enters him as John B. DuBay, age 42, lumberman, having real estate valued at $10,000. In his household were his wife Madeline, six children, one woman and seven men. Neither he nor his wife could read or write, but when a post office was established at his trading post in 1851, DuBay became the first postmaster.

When lumbermen and their teams of laborers arrived in the area, they found wood as the only material for building houses, making fires for cooking and heating water, and constructing some rough furniture. Before long suppliers of other necessary goods arrived in the area with foodstuff, clothes and other commodities.

The 1850 census lists two merchants in Plover and four in the town of Stevens Point. They probably owned and operated general stores. The first issue of The Wisconsin Pinery, the newspaper of Stevens Point, was published on January 14, 1853. It contained four advertisements for stores--none of which were owned by any of the merchants listed in the 1850 census. The four stores were general stores which would now be designated as country stores. One advertised: "Groceries and provisions, glass, nails, crockery and hardware, ready-made clothing, pails, tubs, brooms and other articles too numerous to mention which will be sold at wholesale and retail for prices lower than ever offered in The Wisconsin Pinery, for cash only." Merril and Slothower advertised "Dry goods, ready-made clothing, boots, shoes, pork, cornmeal, buckwheat flour, oats and beans, groceries, hardware, glass, flour, bacon, beef, lard for cash, lumber and shingles." Further, it indicated: "The highest price in cash paid for produce, lumber and shingles at the Barter Store Merril and Slothower." Briggs and Curtis had "Staple dry goods and groceries," whereas John Morrison & Co. were "Wholesale and retail dealers in dry goods, boots, shoes, drugs and medicines, groceries and provisions, hardware and stoves."

There were no specialty stores in Portage County, but the paper advertised a bookstore and a drug and medicine store in Milwaukee. Advertisements indicated what was sold in the outdoor market: pork, beef, flour, cornmeal, oats, butter, lard, salt, hides, shingles, seasoned boards, laths, souvenirs, land warrants, beans, dried apples, cranberries. No mention was made of milk, cheese, eggs, vegetables (except dry beans), and the only fruit mentioned were dried apples.

Two-and-a-half years later, the second newspaper in the county was published: the Plover Herald, No. 1, Vol. 1, August 7, 1856. It carried news, editorials, legal notices, but also had a "Business Cards" column which listed four dry goods businesses. There were also attorneys-at-law (Plover was the county seat), a surveyor and land agent, a Daguerrean artist (the name used for photography after Louis Daguerre, the Frenchman who invented photography. The first photos were called dauerreotypes).

With the development of Stevens Point came the addition of more stores and shops. The first platting of the village of Stevens Point was executed in 1847. It consisted of a number of lots, blocks and streets (Main and Clark, west and east; First and Second streets, north and south). At the intersection of Main and Second Street a space was designated as the Public Square, similar to those in the villages of New England. No special use was initially indicated for the open space, but it was intended for public gatherings. However, businesses were established around it, and it soon became a marketplace where farmers came to sell their produce. At present it is still part of the central business district. The September 9, 1854, issue of The Wisconsin Pinery gave the following statistics for buildings. There were 185 buildings in total: 109 dwellings with 126 families; 4 taverns; 9 stores and 23 shops; 4 blacksmith shops; 3 carpenter shops; 2 shops each for tailors, painters, boots and shoes, and wagons; a bakery; a barber shop; a fanning mill; a machine shop; shops for a gunsmith, a tanner and a harness maker; two saloons and two drugstores. The nine stores were probably general stores, but soon specialized stores opened. On September 11, 1856, The Wisconsin Pinery mentions 9 dry goods establishments, 6 provision stores, 2 clothing stores, 3 boots and shoes stores, 2 millineries; a jewelry store; a meat market; a bakery; a hardware store; 6 saloons and 25 lumber dealers.

Many immigrants came as farmers and settled in the county. General stores opened to serve neighborhoods. The census of 1896 gives details of the various stores; their locations indicate that they were clustered close to where people lived so that they could walk to buy commodities.

A map of Portage County published in 1876, carried advertisements for businesses. In villages, general stores were mentioned under the name of "dealers in dry goods," followed by a list of articles: groceries, clothing, hats, boots and shoes, and notions. Some also handled hardware, crockery, glassware, liquors and cigars. There were two such stores in Stevens Point, one in McDill, two in Ellis, one in New Hope, three in Nelsonville, three in Amherst and two in Almond. There were two dealers in general merchandise in Buena Vista. In addition there were a number of specialized stores: manufacturers and dealers in boots and shoes in Stevens Point, Nelsonville, Amherst and Almond; a wholesale dealer in wines, liquors, cigars and tobacco; a hardware store dealing in shelf and heavy hardware, iron, stoves, tinware; a jeweler and dealer in jewelry, silverware, watches, clocks; a drug store, dealer in drugs and medicines, paints, oils and glass. In McDill there was a dealer in groceries and confectionery. In Amherst there were two dealers in agricultural implements; a furniture dealer; a dealer in fancy goods; a dealer in farm produce; a physician and surgeon who also was a dealer in drugs and medicines and a perfumery for fancy and toilet articles, glass, putty, paints, oils and dye stuffs. Several dealers in lumber and shingles were located near the sawmills.

In 1896, a directory of Portage County was published, with the name and the location of every business in the county. Saloons constituted the most numerous category; there were 71 of them, of which 43 were in Stevens Point (9 on the Public Square and 11 on Main Street); 4 in Polonia; 3 each in Amherst Junction, Custer, Junction City and Plover; 2 in Amherst, Arnott, Ellis and Rosholt; one in Casimir, McDill, Nelsonville and Rosholt. General stores numbered 51: 6 in Stevens Point; 5 in Polonia; 3 each in Amherst, Amherst Junction and Nelsonville; 2 in Almond, Blaine, Buena Vista, Custer, Ellis, Junction City, McDill, Plover, Rosholt and Stockton and, lastly, one in Arnott, Bancroft, Casimir, Eau Pleine, Garfield, Keene, Lanark, Lone Pine, Madely, Meehan and New

Hope. Businesses specializing in foodstuffs, clothing, housewares, hardware, personal articles or services were located in the city or in larger villages. In Stevens Point there were 38 groceries; 12 meat markets; 6 milk dealers; 4 bakeries; 4 fruit and confectionery dealers; 3 florists; 6 drugstores; 6 dry good stores; 3 plumbing, heating and lighting businesses; 6 variety stores; 12 barbershops; 3 laundries, and a number of other stores. Plover, which at that time was no longer an incorporated village, had only two groceries, a meat market, a drugstore, a dealer in furniture and undertaking, a hardware store and two hotels.

Relative to transportation (horses, carriages and wagons), there were in Stevens Point four livery and boarding stables, two horseshoers, three harness makers, eight blacksmiths and wagon makers and 20 others in the villages. Relative to agriculture (dealers in agriculture implements, pumps and windmills), two were in Stevens Point, and eight in the villages. There was one flour and feedmill in Stevens Point, five in the villages.

In addition to the two saloons and three general stores already mentioned, Amherst had three general commission merchants; three barbershops; three blacksmith and wagon makers; two drugstores; two milliners; two livery and boarding stables; a harness maker, a grocery, a meat market, a general hardware store, a milk dealer, a merchant tailor, a jeweler, a restaurant and a hotel; two flour and feedmills, three dealers in agricultural implements, pumps and windmills.

Nelsonville had three general stores, a drugstore, a merchant tailor, a dealer in boots and shoes, a furniture dealer, a general hardware store, a flour and feedmill, a harness maker and a saloon.

Almond had two general stores; two dealers in agricultural implements, pumps and windmills; two blacksmiths and/or wagon makers; a milliner; a harness maker; a hotel and no saloon.

Amherst Junction and Junction City owed their development to their location at the junction of two railroads where travelers transferred from one train to another. Amherst Junction had three saloons, three general commission merchants; three general stores; two hotels; a restaurant; a barbershop; a meat market; a livery and boarding stable; a milliner; a blacksmith and/or wagon maker.

In addition to the two saloons and two general stores already mentioned, Plover had a general commission merchant, two groceries, a meat market, a drugstore, a jeweler and/or optician, a general hardware store a furniture dealer a barbershop, a laundry, a livery and boarding stable, two hotels or boarding houses, and a blacksmith and/or wagonmaker.

Other than saloons and general stores, there were very few businesses in other villages and towns: a flour and feedmill in Bancroft and another in Rosholt; a general merchandise dealer in Arnott; a grocery in Hull and another in Rosholt; a merchant tailor in Ellis; a dealer in agricultural implements in New Hope and another in Rosholt and blacksmiths and/or wagonmakers in Alban, Buena Vista, Keene, Meehan, New Hope, Peru and Polonia (four in this village).

A Stevens Point Community Directory was published in 1931, with 170 headings in the business sections. Among the headings, livery and boarding stables, wagon makers and horseshoers did not appear. However, there were still three blacksmiths and two harness shops. With the development of technology, new types of businesses emerged. There were nine automobile and truck dealers, 38 automobile service shops and 31 filling stations, seven electric appliance dealers, nine radio apparatus dealers and one typewriter dealer. There were five general stores, one of which was J.C. Penney. Stores selling foodstuff were numerous: 2 wholesale groceries and 53 retail groceries; 4 bakeries; 15 meat dealers; 6 dealers in dairy products; 2 fruit dealers; 2 produce dealers and 2 fish dealers. There was a "five and dime" store; 7 dry goods stores and several others offering clothes, shoes, house furnishings and furniture; 8 drug stores; 17 barbers; 7 beauty parlors; 9 hotels and 28 restaurants. Since Prohibition was in force, there were no saloons, taverns or bars, but there were 42 dealers in soft drinks (some called soft drink parlors). Most businesses were in the center of the city, on the Public Square and around it, or in the area between Clark, First, College streets and Strongs Avenue. Located on the Square were ten soft drink dealers, a general store, four groceries, two leather goods stores, two shoe repair stores, a fish dealer, a produce dealer, a tailor, a paint and wallpaper store, a barber, a chiropractor, two restaurants, a rooming house, a finance company and an insurance company. On Main Street there were two variety stores, an optician, a dealer in art goods and one in musical and office supplies.

As the population increased and use of the automobile became more frequent, shopping centers were built away from the downtown area. They were located near a major thoroughfare offering ample and well-lighted parking space. Shopping centers generally started with one major department store and a few other businesses, specialty stores, food stores, service stations, restaurants, taverns and other services.

In Stevens Point the first shopping center was Northpoint Center on Division Street (now Business 51 which was formerly part of U.S. Highway 51 that now bypasses the city). The main store was Tempo and, next to it, a drugstore and filling station which were joined by other businesses. Another shopping center started on the southside (also located on Business 51). A third shopping center was built on U.S. Highway 10, partly in Park Ridge and partly in Stevens Point. There were two stores responsible for the third center--a drugstore and a supermarket.

Other shopping centers established, with parking lots adjacent to Business 51, were Springville Shopping Center, Golden Sands Shopping Center, the Village Plaza in Plover, the Whiting Center, the Church Street Station and the University Plaza next to the North Point Shopping Center. Some were begun by individual owners or companies, others by developers who own the buildings and rent them.

The development of refrigeration, the introduction of frozen foods and the canning and packaging of bread, milk, fruit juices, eggs, etc., were all reasons for the origin of supermarkets. These businesses were also called foodliners, they occupied a wide space with some refrigerated closets and tubs. Their quantity and convenience allowed customers to buy supplies for several days, or even weeks, instead of making daily trips to the store. Supermarkets were frequented not only by city dwellers, but also by people from farms and villages.

Until the 1920s, stores were usually owned and managed by independent persons or families. Consequently they were usually small. With the coming of department stores and supermarkets, owners were usually not individuals, they were companies. The store might even be one chain of stores with headquarters in another city.

A development which took place after World War II was self-service, with cashiers at the stores' exits. Most large stores adopted this system.

A new concept developed in the 60s. Malls were organized by a company, often with the support of a municipality or business association. Malls consisted of one or two alleys and, at each end, an anchor store which was usually a large department store. In between were other stores and some services (banks, insurance agencies, physician's office, travel agency, etc.). They were located at places recommended by a consultant service. The city of Wisconsin Rapids opened the first mall in the region, and was followed by malls in Wausau and Marshfield. In 1971, a developer announced that a shopping center would be coming to Stevens Point, with H.C. Prange as the main department store. It was to be located at the east end of Main Street. Later, in 1978, at the same location a sign announced the future "Soo Mall" which would have Shopko and an IGA supermarket as anchor stores with 40 other stores and services. However, the project never materialized. Another mall was suggested in 1978. For the downtown area, but neither the Downtown Business Association nor the Central Wisconsin Chamber of Commerce would support it because of the impact on local taxes estimated in the first proposals. Finally, after long discussions, it was concluded that a mall was a necessity for the economic growth of Stevens Point. Recommendations were made that the mall be built downtown to revitalize that area and stop what many looked upon as downtown "decay." The plan was eventually accepted and the mall was planned and built. The mall is enclosed to attract shoppers all year round and during all kinds of weather. The two anchor stores at each end of a main alley are J.C. Penney, headquartered in New York City, and Shopko, with headquarters in Green Bay. It was expected that 60 stores would open along the alley. A third anchor store was planned at the north side, but has not materialized.

In 1984, during discussions of the downtown mall, a developer opened a mall in Plover. It is not of the mold established as the base rule of malls by developers, architects and economists. It has no anchor stores. It is in the form of a cross with wide alleys and entrances at each of the four ends of the mall, and it is surrounded by parking space. It is meant to attract the people of Plover and of Stevens Point, but also the farmers and villagers of the region. The mall, called Manufacturers Direct Mall & Outlet Center, has about 40 stores featuring low prices and selling wares direct from factories. The outlet is located in the town of Plover close to U.S. Highway 51. The majority of stores were rented almost from the start. Most were factoty direct outlets, but others were not. Despite the Plover Mall, The Center Point Mall was established and officially opened in October of 1985. Two anchor stores and eight retail stores were the starting tenants. A few more tenants were expected after Christmas. The third anchor store and the remaining small stores were expected to be filled within two years.

The demise of many small businesses came about as department stores and supermarkets offered wide choices and large quantities of merchandise at prices lower than small businesses could afford to offer. Younger owners of small businesses found other occupations, and older owners closed their shops and retired. As competition grew, stores that were depending on large companies (like A & P, Tempo, Woolworth,

Montgomery Ward and Ben Franklin) simply closed their Stevens Point branches.

In Stevens Point, general stores were the first to disappear. They were replaced by stores specializing in foods, meat, bakery products, clothing, household articles, hardware and furniture. Most of the food stores became groceries; their number increased as the population increased. In 1895 there were 6 general stores, 38 groceries, 12 meat markets, 6 milk dealers, 4 bakeries and 4 fruit and confectionery dealers. In 1923 a business directory listed 46 grocers and 36 dealers in other food kinds (some of which were among the grocers). In 1937 the peak of retail food stores was reached. There were 57 groceries (some calling themselves supermarkets), 7 meat markets (not connected with groceries), 6 bakeries, 12 dairies and 1 dealer in fruit.

About 20 years ago, some filling stations began selling various articles, such as candies, cigarettes and soft drinks. They expanded into groceries and became miniature supermarkets. At the present writing there are only ten different general food stores in Stevens Point. Most qualify as supermarkets. Some even sell beer, wine and liquor. There is one bakery with three retail stores in the city, and one specialty store selling cheese, wine, coffee, tea and sausage. There is no longer a store exclusively selling meat, dairy products, fish, fruits or vegetables.

Originally, drug stores sold only drugs and medicines, but since they were the only stores open on Sundays, they were licensed to sell other articles, including newspapers, cigarettes and candies. Four drugstores had advertisements in the 1923 city directory. The Prescription Store advertised school books and supplies, toilet articles, candy and cigars. Another drugstore called itself "the best equipped store in the State of Wisconsin" adding "Prescriptions carefully compounded by graduate pharmacists....and Kodak photo chemicals, developing and printing." The third indicated "Service first - quality always: groceries, drugs, books, stationary, wallpapers, paints, oils, varnishes, photographic supplies." The fourth had "Drugs and drug sundries; dealers in `the new Edison, the phonograph with a soul'." Some drugstores developed into variety stores and giftshops, and some had counters to serve ice-creams, soft drinks, coffee, snacks and lunches. Some had a section as a coffee shop with booths. Supermarkets and department stores started to sell drugs and eventually, had full-fledged pharmacy sections with licensed pharmacists.

While the evolution of retail trade was taking place in Stevens Point, the villages and small centers also changed. A number of specialty stores opened in villages. In 1964 Amherst had a bakery, a clothing store, a florist, two grocers, two hardware stores, a ladies ready-to-wear store, a television and radio shop and an automobile dealer. In Amherst Junction there was a grocer and a general merchandise store. Nelsonville had two grocers, a dairy and a dealer in farm equipment. Almond had a dairy, a drugstore, a dealer in farm equipment, two grocers, a dealer in produce, a hardware store and a junk dealer. Bancroft had a grocery and a dairy. Junction City had an automobile dealer, a dairy, a grocer, a meat market and two dealers in farm equipment. Plover had a bakery, two grocers, a fruit dealer, a gunshop, a junk dealer, a liquor store, produce store and a television dealer. In Rosholt were two automobile dealers, a bakery, a druggist, a general merchandise store, two dealers in farm equipment, four grocers and two produce stores. In Polonia were two grocers, an automobile dealer and a television dealer. Kellner had a grocery and a hardware store. Groceries or general merchandise stores were found in Arnott, Blaine, Custer, Garfield, Liberty Corners (Town of Buena Vista), North Star (Town of Sharon), Peru and Spiritland (Town of Almond).

Stevens Point had approximately 150 retail stores and dealers in 1964. Among the stores and retailers there were six department stores, five variety stores (five & dime) and an array of specialized businesses (such as cameras, church goods, confectionery, florists, fruit, furniture, furs, jewelry, leather goods, millinery, music and musical instruments, newsstand, office supplies, sporting goods, watches and clocks).

Since then department stores and supermarkets have increased in size and offer a wider variety of articles. Self service has become the norm in all large stores. With part-time employees working in shifts, larger stores are open evenings and Saturdays-- several are open 24 hours and on Sundays. Most are part of large chains and can offer merchandise at lower prices than can the small stores. Shopko, with headquarters in Green Bay, has approximately 70 stores in 12 states--from Michigan to Nevada and Washington. IGA, based in Stevens Point, has 31 stores--all in Wisconsin. Another large store is Mills Fleet Farm on Highway 10 East, a branch of an Appleton business with 26 stores in Wisconsin and Minnesota. In 1988 Fleet Farm moved to a new location (near the old one) occupying 130,000 square feet. The business caters to farmers and customers interested in building or remodeling their homes. Store items cover almost all customer needs, with sections in automotive supplies, building materials, electrical and plumbing supplies, tools, household and kitchen appliances, gardening needs and clothing, and items for fishing, hunting, camping and other sports. It also has a gas station and automotive service center.

Almost two years after the opening of CenterPoint Mall, a new section was added: CenterPoint on Main - The Connection. It has two stories with a number of quality stores in jewelry, women's apparel, home appointments and gifts.

Even though there are large department stores and supermarkets, there is still a place for specialty and quality stores. Some customers want unusual or custom designed items, or they want to discuss their choice with an informed salesperson. Often the client is well known by the owner, manager or employee in a small store; the purchase is at the same time as the visit. A number of specialty shops have articles not found in large stores, or at least without much choice. For instance, books, musical instruments, furniture, photographic equipment, sports equipment and antiques. Specialty shops may sell some articles at a higher price than larger stores but, because of their particular attention to the customers' needs, there will always be a niche for specialty and quality shops in the marketplace.

The 1982 Federal Census of Retail Trade produced a list of retail establishments (including eating and drinking places, and gasoline service stations). The total for the county was 503 establishments, of which 341 were in Stevens Point, 45 in Plover and 117 in the balance of the county. The census categorizes only those establishments with payrolls, operated by an individual or family and having regular employees. Those businesses amount to 204 (excluding eating and drinking places, and gasoline service stations). The figures are as follows for the county, Stevens Point, Plover and the remainder:

	Portage County	Stevens Point	Plover	Rest of county:
Building materials, hardware, paint, glass, garden supplies	28	14	4	10
General merchandise stores: department (w/50+ employees) general (w/-50 employees)	8	5	1	2

Food stores: groceries, 29 20
2 7
bakeries, dairies, meat, fish,
poultry, public market

Automotive dealers: cars, 19 14
1 4
pickups, vans, campers,
motorcycles (excluding gas
service stations)

Apparel, accessory stores: 29 27
1 1
clothing, shoes, custom
tailors, dressmakers

Home furniture including: 22 13
2 7
floor coverings, drapery,
kitchen/table ware, household
appliances, radio, tv,
computers, music stores

Drug stores 9 7
2 0

Miscellaneous: liquor, books, 60 52
3 5
jewelry (watches), leather,
florists, tobacco, news,
optical, religious, trophies,
pet shops

Total

---

businesses
w/regular employees: 204 152
16 36

Since 1982, Stevens Point's trade picture has gone through some important changes: the dedication of CenterPoint Mall CenterPoint on Main, the enlargement of some department stores and supermarkets; the closing of some small stores and the opening of several new specialty shops.

Amherst has been able to keep a number of stores and add some in recent years. At present there are two food stores, a bakery, two television and radio stores, two antiques stores, a hardware store and a feed store. Nelsonville has a large food store selling groceries and meats, and a large business selling agricultural implements. Amherst Junction no longer has any retail stores, but it is near Nelsonville and Amherst, and not far from Stevens Point. Custer has no retail stores, but Arnott, which is not far away, has a general merchandise store. Bancroft has a food store and a hardware store. Almond has a food store and a hardware store, a store selling baits and tackle and an army surplus store that is not regularly open. Kellner has one food store. Junction City no longer has any large food or hardware stores, but just outside town is a small store owned and managed by an individual dealing mostly in hardware and sporting goods and offering a variety of other commodities as well. Polonia no longer has any retail stores, but there are taverns that sell some commodities. In Rosholt there is one food store and two hardware stores. Three more grocery or general stores are outside villages: one called Liberty Corners at the intersection of Highway 54 and County Trunk J; another, the Alban Store, is near the intersection of Highway 66 and County Trunk A, not far from Rosholt and the Blaine Store at the intersection of County Trunks A and I.

Places that had some food stores 20 years ago, such as Garfield, Peru and Keene, no longer have any stores or businesses.

TRANSPORTATION

Before the region was opened for settlement, Indians probably used the Wisconsin River and the Plover River as waterways. They established trails on land, along the rivers and in the interior. At first trails were narrow, only wide enough for foot travel. Eventually trails were widened for the use of horses, and in the era of pioneer settlers, for wagon trains.

It is presumed that Plover was important to Indian travel. The trail following the Wisconsin River met the trail coming from the east. Plover marked the end of the trail coming from the Fox River and the Waupaca River. This trail was the boundary between the Chippewa and the Menominee tribes, and it is doubtful that it would have been used for migrations.

In surveying the land in 1839, Hathaway refers to Plover Portage, and it is only in 1843 that the term Plover Portage appears in a document. As a portage, Plover was a good site, a trail connecting the lower Wisconsin with the Upper Wisconsin, avoiding various rapids. It is here that a few houses and a tavern were the first white settlements of the region. If Plover had been at a favorable spot on Indian trails, it would have been there that John Baptiste DuBay in 1830 would have established his trading post for the American Fur Company. He chose, however, another Indian trail (at present under the water of Lake DuBay), the first point north of Petenwell Rock in Adams County where Indians could cross the stream by foot or on horseback.

When the country was first explored by the white man, travelers also used the rivers and the Indian trails. The south-north trail on the left bank of the Wisconsin River was given official recognition in 1841 when the territorial legislature authorized a road from Fort Winnebago (the present city of Portage) to Plover Rapids. It was called the Pinery Road. Other trails were traced as travelers came to the lumbering country from Green Bay and Milwaukee. Most pioneers who came to Portage County after 1845 passed through Waupaca County and entered in the southwest, into Plover. Others coming from Portage came through Montello, which is now Almond. There was still another road of entry, from Lake Winnebago and the Wolf River through Weyauwega to Plover. Though this road was little used at first, it became important in 1853 when Gill's Landing was established on the Wolf River, making access to Plover easier. In 1854, a stage line was organized between Gill's Landing and Plover.

A scheduled stage line had already been established in 1847 from Fort Winnebago to Plover via the Pinery Road. At first, stages came to Plover, but by 1857 Stevens Point was listed as the termination point of daily U.S. mail routes from Weyauwega, Berlin and Portage City, and tri-weeky mail from Grand Rapids. Among early roads was the road from Plover to Wausau, the forerunner of Highway 51; and Airline Road (possibly Urline Road) still partly existing which connected Stevens Point with the road from Plover to Berlin, at Keene. Though territorial legislature had authorized construction of the roads, it appears the roads existed before any governmental action had been taken.

As the village of Centralia in Wood County grew in importance, it became the shopping center of the west bank of the Wisconsin River. Though there was no bridge across the stream, even before 1850, a ferry service had been organized near the mouth of Mill Creek. Eventually Centralia was joined with Grand Rapids to form Wisconsin Rapids. Steamers left daily from a dock at the foot of Main Street to sail via the Wisconsin River to Mosinee.

When the railroad in Wisconsin, the line from Milwaukee to Wauwatosa, began to operate in 1850, it was considered to be of great importance to Wisconsin. It was expected that soon there would be railroad lines all over the state, encouraging settlement of the territory, stimulating the exchange of goods and developing the economy.

"The Pinery", the first Stevens Point newspaper which began publication in January 1853, carried an advertisement with a picture of a train and the text:

GREAT EXCITEMENT!!
RAILROAD COMING!!

General Depot at Rood's where will be found constantly on hand a large assortment of groceries and provisions, glass, nails, crockery and hardware, ready-made clothing, pails, tubs, brooms and other articles too numerous to mention which will be sold at wholesale and retail prices lower than ever offered in the Pinery, for cash only.

-Stevens Point, January 10, 1853.

This advertisement was published in every issue of the paper until August 17 of that same year. Thereafter the headline was changed to: A NEW ERA IN THE HISTORY OF STEVENS POINT, but there was no mention of the railroad. Almost every issue of the newspaper had articles on railroads, extensions of railroad lines, invitations to buy stock in railroad companies (particularly the Milwaukee, Horicon, Berlin and Stevens Point Railroad Company; and the Green Bay and Minnesota Railroad). At the same time however, there were advertisements of stagecoach schedules from Plover and Stevens Point to Gill's Landing, and schedules for steamship lines from Oshkosh and Fond du Lac to Gill's Landing.

In January 1854, the Milwaukee to Mississippi Railroad completed the line to Stoughton, thus providing pleasant travel from Stevens Point to Milwaukee. A person could drive in eight hours from Stevens Point to Madison, stay in a hotel there; the next day take a two-hour trip by stagecoach to Stoughton and then take a seat in a warm, comfortable railroad car and arrive in Milwaukee in time for tea.

There was some debate that the Capitol should not remain in Madison, but should be moved to the center of the state. The idea was published in the Sheboygan Journal, and the place suggested was Stevens Point. The city was in the geographic center of the state, in a pleasant location, with healthy climate, distant from large cities where political influence might be exerted and with the potential of becoming the main junction of railroads in Wisconsin. The Pinery was happy to support the suggestion that had been on the mind of local people. At the time, Stevens Point was not incorporated. According to The Pinery, January 14, 1854, Stevens Point consisted of 91 houses, 10 or more stores, 20 workshops, five taverns, sawmills, school houses, churches and a population of nearly 700.

On April 19, 1855, The Pinery published an extra feature, a map of Wisconsin showing the routes of the principal railroads chartered and under construction. Principle railroad lines were to converge in Stevens Point:

Illinois-Beloit-Madison-Portage City-Plover

Whitewater-Jefferson-Columbus-Montello-Plover Milwaukee-Horicon-Waupun-Berlin-Plover

Manitowoc-Menasha-Weyauwega-Waupaca Falls

Green Bay (direct)

Ontonagan, Michigan(direct)

They represented five companies:

Madison-Lake Superior R.R.

Central Wisconsin R.R.

Milwaukee & Horicon R.R. (later Milwaukee-Horicon-Berlin-Stevens Point R.R.)

Green Bay & Minnesota R.R.

The Milwaukee-Horicon-Waupan-Berlin-Stevens Point line was the first to reach Stevens Point, and since the line passed through the towns of Belmont, Almond, Pine Grove and Plover, farmers of those towns bought stock in the company. The year 1857 was marred by rampant speculation in American railroads and land. Many companies went bankrupt. The Milwaukee-Horicon-Waupan-Berlin-Stevens Point Line was built to Berlin, but not extended any further. It was only in 1971 that the first train from the Wisconsin Central Railroad Line arrived in Stevens Point from Menasha. The following year the Wisconsin Central Line was extended to Marshfield, and the Green Bay and Lake Pepin Line to Merrillan.

In 1873, the Wisconsin Valley Railroad arrived in Junction City from Tomah; the following year it was extended to Wausau. In 1875, the Wisconsin Central began service on the Portage branch; first from Stevens Point to Hancock, and later to Portage City and Madison. In 1901, the Chicago and Northwestern Railroad built a line from Princeton to Marshfield, passing through Almond, Bancroft and Kellner in Portage County. The last line to be constructed, built in 1903 by and for the benefit of a lumber company, connected Rosholt to Eland in Marathon County.

Steamships and stagecoaches ceased their activity when they were no longer able to compete with the railroads.

Railroads built to serve such existing centers as Stevens Point, Plover, Amherst and Almond, also helped those communities to develop. Other centers started at railroad junctions such as Amherst Junction and Junction City, or along railroad lines such as Bancroft, Custer, Stockton, Arnott and Kellner. After a spur line built by the Chicago and Northwestern Railroad Company was laid out from Eland to Rosholt, the area became a boom town.

The phenomenal success of the railroads after 1870 caused highways to become secondary in importance. The three decades between 1870 and 1900 proved to be a "dark age" in the development and maintenance of rural highways in the state. As far as highway traffic was concerned, it was limited to horse-drawn vehicles traveling from home to meeting places: markets, mills, railroad stations and churches.

However in 1871, Dr. Carhart of Racine designed and operated the first light, self-propelled highway vehicle in the United States, and probably in the world. The legislature, impressed with Carhart's invention, offered a $10,000 reward for the invention of a self-propelled vehicle which would successfully run over 200 miles of public highway. The contest, the first automobile road race in the United States, was held in July of 1878. Seven registrants filed for the contest, but only two started and both had two-cylinder steam engines. The vehicles were named "the Green Bay" and "the Oshkosh." "The Green Bay," because of superior speed, led the race to a point 20 miles from Madison, but did not finish because of mechanical difficulties. "The Oshkosh" finished the race in 33 hours and 27 minutes with an average speed of slightly more than six miles per hour. The legislature offered a prize of only $5,000 since the vehicle did not prove to be a cheap and practical substitute for horses or other animals on the highway and farm. Other types of automobiles were developed and, in 1899, there were 3,700 automobiles manufactured in the United States. The number increased to 356,000 in 1912, but even that was not enough to be a threat to horse-drawn carts and steam trains.

In 1905, a count of 1,492 automobiles and other motorized vehicles was registered in Wisconsin. Their numbers increased slowly and 6,152 were registered in 1911. That year the State Highway Commission was established and 28,638 vehicles were registered in 1912. This doubling of registrations in one year was probably due to the fact that until 1911, registration was not mandatory. Registration numbers continued to grow: in 1918, more than 200,000, and one million in 1942. For the year 1985-1986, the total was 3,404,603 and distributed as follows:

Automobiles	2,349,441
Trucks	818,102
Trailers, Semitrailers	75,968
Motorhomes	72,195
Busses	11,712
Motorcycles	174,369
Mopeds	22,275
Municipals (all kinds)	57,166

In 1920, the figures published for Portage County were:

Automobiles	3,356
Trucks	106
Motorcycles	70

Unfortunately, statistics according to counties are no longer published. Total motor vehicles in Portage County at this writing is approximately 34,000.

The legislature of 1917 directed the State Highway Commission to establish a highway system not in excess of 5,000 miles, interconnecting every county seat and every city with a population of 15,000 or more. In 1919, the network was expanded, adding 2,500 miles to the original highway system. In 1923, the highway network was again extended, not to exceed 10,000 miles.

Wisconsin led the nation in marking and numbering the state trunk highways; other states followed. It soon became apparent that the entire United States required a standard design by which interstate highways would be marked and numbered continuously without regard to state boundaries. Such a system was selected by representatives of the United States Bureau of Public Roads and of the state highway departments in Washington D.C., 1925. The representative logo of the system was a shield with the upper portion naming the state in which the highway section was located and below, the abbreviation U.S. and the highway number. Principal east and west highways were designated with even numbers, the smallest number starting at the north. Principal north and south highways were indicated by odd numbers, the smallest number at the east. Intermediate highways contained some of the same digits as U.S. highways that linked them. For example, U.S. 41 from Milwaukee to Green Bay through Fond du Lac and Appleton; U.S. 141 from Milwaukee to Green Bay through Sheboygan and Manitowoc. Even though highways are marked with the U.S. logo, there is no corresponding implication that U.S. funds are available for their improvement.

Portage County is crossed by two U.S. highways which intersect in Stevens Point: U.S. Hwy 51 which runs from Hurley to New Orleans in an almost straight line, and by Hwy 10 which in Wisconsin runs in a near straight line from Manitowoc to Minnesota near St. Paul (however the highway is not continuous from the Atlantic to the Pacific, missing links in several states). Where federal systems incorporated existing state highways, the federal number overrode the state number--i.e. U.S. Hwy 51 replaced old Wisconsin 10 and U.S. 10 replaced old Wisconsin 18. Some numbers did not change however, and in Portage County there are stretches of several highways that still hold the state number.

In 1945, mileage of highways in Wisconsin was:

U.S. 10 38.26 miles (61.54 km)

Wis. 22 1.64 miles (2.25 km), Wautoma to Waupaca
Wis. 34 13.10 miles (21.08 km) Wis Rapids to Hwy 51 via Junction City
Wis. 49 6.81 miles (10.6 km) Waupaca to Hwy 29 via Iola, Elderon
U.S. 51 34.39 miles (55.34 km)
Wis. 54 23.12 miles (37.2 km) Wis Rapids to Waupaca via Plover
Wis. 66 18.60 miles (29.93 km) Stevens Point to Hwy 49 via Rosholt
Wis. 73 1.93 miles (31.06 km) Wautoma to Wis Rapids via Plainfield
Wis. 161 5.53 miles (8.90 km) Hwy 10 to Hwy 22 via Nelsonville

(Between 1945 and 1987, only 9.16 miles (14.6 km) were added to this network.)

No section of the interstate system passes through Portage County--the nearest is the stretch common to Interstates 90 (Illinois, Wisconsin, Iowa) and 94 (Illinois, Wisconsin, Minnesota) which crosses U.S. Hwy 51 near Madison.

A system of county trunk highways, designated by letters, was establish by the State Highway Commission in 1925. Wisconsin highways are divided into several categories. Those in Portage County are distributed as follows, as of January 1, 1987:

State trunks including federal		54.54 miles (248.80 km)
County trunks		443.20 miles (713.59 km)
Local roads:	Town	1,003.83 miles (1,615.16 km)
	City	102.74 (165.30 km)
	Village	101.90 (163.95 km)
	Total	1,208.47 (1,944.28 km)

Other roads: parks and forests 4.44 (7.14 km)

Total county roads mileage: 1,810.65 (2,913.33 km)

In 1973, the State Legislature created the Rustic Road System in an effort to preserve what remains of Wisconsin's scenic, lightly travelled country roads for hiking, biking and other recreational enjoyment. There is such a road in the county, in the town of Lanark. It is marked by a sign R-17, starts from Highway 10, runs for 2.7 miles and ends at County Trunk T.

Due to the number of streams, there are 132 road bridges: one over the Wisconsin River for Hwy. 10, 58 for other state trunk highways, 33 for county highways and 40 for local roads. Eight are overpasses on U.S. Hwy 51 which run east of downtown Stevens Point to avoid downtown traffic.

The first suggestion of a bridge over the Wisconsin River appeared in the Stevens Point newspaper "The Pinery" on February 13, 1863. Construction started in May 1866, but when the wooden bridge was near completion a heavy storm lifted the structure and dumped it in the stream. Construction was resumed and on October 11, 1867 the bridge was completed. Because of the high cost of maintaining a wooden bridge, the city council voted to replace it with an iron bridge having sidewalks on both sides. The bridge was constructed and tested in Milwaukee before being shipped to Stevens Point. It was in place on November 10, 1877. On Memorial Day 1923, disaster struck, fire broke out and the entire structure was destroyed. Ferry service was immediately established, but a new bridge was urgently needed. A temporary bridge was built on July 18, 1923 and replaced with a new bridge on January 1, 1926.

Among other bridges, one that deserves mention is the suspension bridge, a rare type in the country, which spans the Little Eau Claire River, north of County Trunk E, built in the 1870s or 1880s. At one time a suspension bridge existed over Mill Creek in the town of Linwood, but it was dismantled in recent years and has not been replaced.

The Wisconsin Department of Transportation, which replaced the State Highway Commission, periodically counts traffic on main roads and on some village and city streets, publishing the average values. The volume of traffic is carefully studied. Automatic recorders have been fixed in Coloma, Waushara County, on U.S. Hwy 51 and in Sheridan, Waupaca County on Hwy 10. Though there are no automatic recorders in Portage County, some conclusions can be drawn from the ones in nearby Waushara and Waupaca counties. In 1985 at Coloma, the annual average daily traffic amounted to 3,720 vehicles, with a maximum of 4,984 on Fridays and a minimum of 2,887 on Tuesdays. Ten percent more traffic was registered for each hour between 2 p.m. and 5 p.m. Traffic decreased before and after those peak hours to reach a low of less than one percent hourly between 1 a.m. and 4 a.m.

At Sheridan, the annual average daily traffic of 5,678 vehicles, with a maximum of 7,304 on Saturdays and a minimum of 4,901 on Tuesdays. Seasonal difference was also evident: a maximum of 8,803 vehicles an hour on Sundays in July, and a low of 3,747 on Saturdays, in January. The hourly average followed the same trend as at Coloma. In 1985 however, the figures of Coloma do not show a true picture since that year construction decreased normal traffic. The following figures show a progression since 1965, except for 1985:

Average annual daily traffic volume:	Coloma	Sheridan
1965	2,078	2,042
1970	3,412	3,167
1975	3,984	3,959
1980	4,296	4,572
1985	3,720	5,678

(figures in the report indicate the preceding year named)

Both U.S. highways in the above report seem to have about the same traffic volume.

Estimates based on counts on certain days have also been published for certain streets in cities and villages. In 1984 in Stevens Point, the maximum vehicles passing during a 24-hour period was 24,890 on Division Street between Main and Clark streets; 18,290 on Division Street between Main Street and Fourth Avenue; 17,550 at Post Street on the McDill Pond bridge; and 13,730 crossing the bridge over the Wisconsin River.

From the beginning of the century, automobiles and trucks offered many advantages over the horse or ox drawn carts. Those who could afford cars, soon owned them.

Competition arose between automobiles and trains. Previously, trains were the most common transportation for people and goods. With the increase of autos, however, the number of train passengers declined. With the increased manufacture of trucks and vans and the improvement of roads, shipment of goods via train also declined. Since 1930, railroads have seen a decrease in the number of passenger trains, and eventually passenger service in may areas was discontinued. Where service was not profitable, service on a line was abandoned, but the procedure for abandoning service, either for passengers or cargo, was so complicated and slow that it took years before a notable decline was measurable in high losses. It was only then that a formal request for abandonment was filed with the government, followed by the suspension of passenger service or the abandonment of an entire line.

Bus services that paralleled railroad lines were also an important factor in eroding the railway industry. In the 1870s, the Green Bay Railroad Company carried about 50,000 passengers a year. With the wave of immigration in the 1880s, the number doubled. 1915 was the peak year, when passenger count reached 310,000. By 1941, the number of train passengers had dwindled to less than 1,000. In 1949 the service of passenger trains was discontinued. On the Soo Line between Chicago, Stevens Point and Minneapolis and Superior, the passenger service was discontinued in 1967. The last passenger line through Portage County was the Milwaukee Road, from New Lisbon to Junction City and Wausau. It was discontinued in 1970.

Lumber, minerals and grain were the main goods transported long distances via railway lines in Portage County. When the Soo Line from Stevens Point to Portage was built, the intention was that it be part of a line from Lake Superior to Chicago. By starting in Stevens Point and ending in Portage, the Soo Line remained a regional line going through rural country that did not offer a sufficient amount of produce from the land, nor from the nearby industries to make a profit, and it was abandoned in 1946. The second line to go was from Rosholt to Eland. Since Rosholt was a logging center, it was plausible that a line be built for the lumber industry. But as the Rosholt line was only a branch line, trains could not compete with trucks which delivered freight faster.

Other sections that were abandoned were two sections of the Chicago and Northwestern Railroad. The first, from Wisconsin Rapids to Bancroft, was abandoned in 1975 and the second, from Bancroft to Ripon, in 1981.

In 1967, the Green Bay and Western Railroad Company received an offer from the Burlington Northern Railroad. Subsequently, Chicago & Northwestern, the Milwaukee Road and the Soo Line made a counter offer, indicating they would abandon 155 miles out of a total 250 miles of the Green Bay route (especially the lines from Manawa to Plover and from Wisconsin Rapids to Winona). Neither offer was accepted.

In 1987 only two railroad lines remained in Portage County: the Green Bay route for freight, passing straight across the county through Plover; the Wisconsin Central Railroad line from Manitowoc to Minneapolis,

Superior through Stevens Point, New Lisbon to Wausau and Bradley through Junction City.

Over the years, all railroad companies that began operation in Portage County have undergone changes through mergers, sales and acquisitions. Wisconsin Central, the first in the county, came out of the merger in 1869 of Winnebago and Lake Superior Railroad and the Portage and Superior Railroad. Minneapolis, Sault Ste. Marie Atlantic Railway, which was formed in 1883 by a group of Minneapolis businessmen to provide a route to the east independent of Chicago, gave way in 1909 to the Minneapolis, St. Paul and Sault Ste. Marie Railroad Company. The latter company took over operations of Wisconsin Central under the name of Stevens Point Division.

In 1961 the Soo Line was formed through the merger of three railroads: Minneapolis, St. Paul & Sault Ste. Marie Railroad Company; Wisconsin Central Company, and Duluth, South Shore & Atlantic Railroad. The former Wisconsin Central network became the Lakes State Division of the Soo Line (Soo derives its name from the pronunciation of Sault Ste. Marie, terminus of the original line built by Minneapolis, St. Paul & Sault Ste. Marie). On October 11, 1987, Lakes States Division was taken over by Wisconsin Central Ltd. which had acquired it from the Soo Line. The headquarters are in Park Ridge, Illinois but operating headquarters are in Stevens Point.

Wisconsin Valley was acquired by the Milwaukee Road, a company that had been organized in 1847 as the Milwaukee and Waukesha Railroad Company and changed in 1850 to Milwaukee and Mississippi Railroad Company. In 1874 it became Chicago, Milwaukee, St. Paul. In 1927 the Chicago, Milwaukee, St. Paul and Pacific Railroad Company, known as Milwaukee Road, was the result of a consolidation of more than 150 smaller railroad companies. Because of financial difficulties, the Milwaukee Road was acquired by the Soo Line. The line from New Lisbon to Junction City and Wausau became part of New Wisconsin Central Ltd. Green Bay and Lake Pepin Railroad became Green Bay and Western Railroad, known as the Green Bay Route.

As highways were improved for motorized vehicles, bus lines were established. Some duplicated railroad lines, others served villages where there was no rail service. The 1958 Centennial Book of Stevens Point states that the city was served by 38 inbound and outbound busses daily to: Chicago, Milwaukee, St. Paul, Minneapolis, Duluth, Green Bay, Madison and points intermediate and beyond. Service was provided by

Northland Greyhound, Central Valley and Seymour Bus lines. As the number of private cars and other motor vehicles increased, the number of bus passengers declined (especially in local traffic).

Greyhound is the principal carrier, with three daily services to Milwaukee and Chicago; daily service to Madison and Chicago; and daily service to Eau Claire, Wausau and Minocqua. A second company, Trailways, operates one daily service from Duluth to Stevens Point, Appleton, Milwaukee and Chicago. A third company, Hiawatha, has a daily bus from Stevens Point to LaCrosse. Hiawatha, which operated lines in Minnesota and western Wisconsin, took over the route from Tomah to Stevens Point which had been handled by Seymour Transfer Line, Inc. (Huettl Bus Lines). At one time these lines had two daily busses from Green Bay to Stevens Point via Seymour, Appleton and New London; and one bus drove to Tomah and back, except on Sundays.

As the use of private cars increased, bus lines lost passengers and service deteriorated. In an attempt to increase passengers, some lines began to pass through towns that had no service; but results were longer rides for people traveling long distances.

Another means of transportation, the airplane, has hurt transportation by railroads, and to a lesser extent, bus lines. There is reference in 1929 of a Stevens Point man who bought an airplane and started his own airfield in the town of Hull. Another private airfield developed in the town of Plover in 1936. Owners gave flight instructions, offered chartered flights and provided crop dusting service. When World War II broke out, a pilot training program was started at Central State Teachers College. Subsequently, the city of Stevens Point made the decision to build an airport.

The Stevens Point airport was dedicated in 1942 and became a focal point in pre-flight training for army and navy cadets, as well as civilian pilots. After the war, the airport manager maintained a DC-3 and daily flights were made to northern and central parts of the state. The first permanent airline was established in 1948 between Stevens Point and Clintonville. Eventually this became part of North Central Airlines which served Wisconsin cities and extended to Chicago, Minnesota and North Dakota. In 1958 North Central had 11 flights in and out of Stevens Point. As North Central Airlines developed, runways of the municipal airport were expanded to accommodate prop-jets. In 1967 the first prop-jet, an AD-9, began service to Stevens Point, the only airport in Central Wisconsin with a runway long enough for such a plane.

Since Wausau, Wisconsin Rapids and Marshfield also had airports, the three cities and Stevens Point decided to establish a common airport. In addition to North Central, other airlines began to use the new airport. In 1969, the same year service began, Midstate (an airline based in Marshfield) started commuter service between Stevens Point, Milwaukee and Chicago. Fare from Stevens Point to Chicago was $26. Service at Stevens Point Municipal Airport was discontinued in 1978, since the company could not generate enough business to make a stop profitable. The Municipal Airport continued to offer flight training, aircraft maintenance, airport management and landing of private planes. Several hangers are available for local residents to keep their private planes.

North Central Airline merged with Southern Airline and the company took the name of Republic Airline. Later Northwest Airline acquired Republic, forming one of the major airlines of the United States. The company has two hubs, Minneapolis and Detroit. At first Northwest had only connections from Central Wisconsin Airport to Minneapolis. Recently it re-established a flight from CWA to Milwaukee three times daily. Four airlines serve CWA: Northwest together with its subsidiary Mesaba Air Link which has seven flights to Minneapolis and three to Milwaukee; United Express, subsidiary of United Airlines, with four flights to Chicago O'Hare (two have a stop at Appleton, one has a stop at Green Bay and the other is a non-stop flight) and one flight to Rhinelander. Midstate has one flight to Chicago Midway Airport. CWA has had a decline in the number of flights and in the number of passengers because the airline has no direct flights to Chicago O'Hare and because of an increase in fares. It now costs $147 for a standard single flight to Chicago O'Hare, so some cost conscious businessmen drive to Milwaukee, Chicago or Minneapolis instead of flying.

In the early 1900's in Stevens Point, a public system of transportation was organized through private initiative with no financial help or material support from the government. It was a loosely-knit, but adequate system for transporting farmers and area growers into the city to carry on their trading, buying and provisional needs. One individual took the initiative to start a public service and in the mid and late 1920's, money was invested in a few motor buses to be used solely for public transportation. There were many customers, especially students traveling to and from elementary schools and high schools. In 1947, more than 580,000 passengers were transported. Although some subsidy was granted from the city, the number of passengers declined to 380,000 in 1959 and in 1968, passengers numbered only 105,000. The reason was largely due to the busing of children to schools by the public school buses. As the owner was losing money, he gave up the operation of the bus service and the city took it over. However, the cost was high and on May 15, 1972, the City Council voted to discontinue the service. A small group of citizens responded by creating a cooperative called PABCO (Point Area Bus Cooperative). It organized a network of five lines: Southside, Southeast, Northside, Westside and Eastside. All of them start downtown on Strongs Avenue at the corner of Main Street. On February 2, 1976, PABCO set a ridership record of 860.

In 1978, the city took over the operation of the transit system and hired its own employees. With local, state and federal subsidies, the Point Transit System has been able to continue bus service, although the number of passengers is very low. Presently there are five routes, all starting at CenterPoint near one entrance to the CenterPoint Mall. The lines are: UWSP/Park Ridge; Northpoint/SPASH; Rice Street/Dixon Street; Water Street/Whiting and Westside. The lines are one way: only Line Four between CenterPoint and Whiting has fares both ways (on lines one, two and three there are two services an hour) and on lines four and five (one hourly service), all starting at 6:45 a.m. and ending at 5:15 p.m. They operate on Saturdays, from Labor Day to Memorial Day. With its present routing and schedule, the bus service has mostly elderly and handicapped customers.

Another means of transportation is the bicycle. For many years there have been bicycle riders, especially young people riding for pleasure or traveling to school. In the early 1970's, news spread that gas was becoming scarce and at the same time there was a popular notion that physical exercises, among them bicycling, was good for a person's health. The result was a boom in the sale and use of bicycles all over the country. In 1974 in Portage County, only Stevens Point, the villages of Plover, Whiting and Park Ridge required a license for bicycles. In 1974 the police issued 7,843 licenses in Stevens Point; 656 in Whiting; 465 in Plover and 275 in Park Ridge--a total of 9,239. The number of bicycles in the county was estimated at 17,000. Now licenses are issued in Stevens Point, the villages of Plover, Whiting and Park Ridge and in the towns of Hull and Plover. The figures are approximate: 8,904 in Stevens Point; 1,546 in the towns of Hull; 1,518 in the village of Plover; 457 in Whiting; 550 in the town of Plover and 320 in Park Ridge, for a total of 13,295. In the county, there are probably 20,000 bicycles. Bikes are used as transportation to work and to school, but many use them for pleasure and for health.

With the development of motorization and the improvement of highways, almost every family in the county had one or more motor vehicles. Private means of transportation were used to get to work, to school, for pleasure, in visiting relatives and for longer trips to Madison, Milwaukee, Chicago and Minneapolis. This resulted in the decline of railroad passenger service, and the consequent discontinuance of service. Bus service also declined. In an effort to attract more passengers, buses served more localities resulting in longer, slower and indirect routes between cities. With the creation of Central Wisconsin Airport, distant from the cities, the curtailing of a number of flights to Milwaukee, Madison, Chicago and an increase in fares caused many travelers to drive to the big cities, instead of flying. Public transportation has in many ways, lost its popularity.

GENERALITIES

Portage County is situated almost in the center of the western half of the northern hemisphere, as it lies between north latitude 44o14' and 44o41' (almost halfway between the Equator and the North Pole) and longitude west 89o13' and 89o51' (almost halfway between the prime meridian of Greenwich and the dateline) longitude 180o. Portage County, close to the center of the state of Wisconsin, is bounded in the south by Waushara and Adams counties, in the east by Waupaca County, in the west by Wood County and in the north by Marathon County. The northeast corner touches Shawano County.

Portage County is in the Great Plains. The only relief is represented by a few outliers of a cuesta, and the irregularities in deposits of material made by glaciers - the moraines. The most important moraine is the terminal moraine of the last Wisconsin glaciation. The highest point is in the town of Stockton. Though not precisely determined by the Geologic Survey, the height is at about 1,300 feet (396 meters) above sea level and the lowest point is where Emmons Creek leaves Portage County at about 918 feet (280 m). The local relief of the county is therefore only 385 feet (116 m). The city of Stevens Point lies between 1,068 and 1,100 feet (322 and 335 m).

According to the Department of Natural Resources, the area of Portage County covers 522,137 acres (211,301 ha) which represents 515,390 acres (208,571 ha) of land and 6,747 acres (2,730 ha) of water. Among the 72 counties in Wisconsin, Portage County ranks 30th in area, whereas in population based on figures of the 1980 census, it ranks 22nd.

When the county was created, its boundaries were based on the U. S. Land Survey System. This system of measuring townships and ranges was used for all land that had not been surveyed before, either by the English, the French or the Spanish. It uses north-south lines (the meridians called the township lines), and east-west lines (the parallels called the range lines). For surveying purposes, a number of regions were determined, each with a principal meridian and a base line (the starting lines for measuring land). Both township and range lines are divided at six mile (9.7 km) intervals. The resulting squares are the congressional townships which theoretically are 36 miles square (93 km2) or 640 acres (259 ha). Starting from the principal meridian, the townships are numbered with the indication of north or south of the base line; the ranges are numbered east and west of the principal meridian. Each congressional township is divided into 36 sections numbered according to a fixed pattern. Each section is divided into quarters, which in turn are also divided into quarters. Each quarter has an area of 40 acres (16 ha) and are often called forties. The forties can be subdivided further into quarters and halves.

In Wisconsin, the U.S. Land Survey uses the 4th principal meridian and the Illinois border as base line so that all townships are north and ranges are either east or west. In Portage County, these areas are all east.

With the U.S. Land survey, any tract of land can be clearly measured and any tract, or even any building, can have its location clearly described. Thus the Public Square in Stevens Point is located in the Northwest quarter of the Northwest quarter of Section 32, Township 24 North, Range 8 East, shortened to NW1/4 NW1/4 Sec. 32, T 24 N, R 8 E.

It is only in theory that each township has an area of 36 miles square (93 km2) and a section one mile square (2.59 km2). Since meridians are lines meeting at the poles and are not parallel, no township or section is square, but is a quadrilateral with the southern side longer than the northern. In order to decrease the discrepancies that occur between the northern and southern range lines, correction lines have been designated: north of them townships and sections are shorter than six miles (9.6 km) and one mile (1.6 km), and north they are exactly those figures. Such a correction line is the Waushara-Portage county line. It may be observed that along that line some of the roads stop, and a few may start on the other side of the line but at a few feet distance.

Portage County is included in the U.S. Land Survey. It consists of 22 townships, namely:

Township 21 North, Ranges 7, 8, 9, 10
Township 22 North, Ranges 7, 8, 9, 10
Township 23 North, Ranges 7, 8, 9, 10
Township 24 North, Ranges 6, 7, 8, 9, 10
Township 25 North, Ranges 6, 7, 8, 9, 10

The southern boundary of the county is a correction line, consequently the sections in Township 21 are almost exactly one mile square (or 640 acres = 2.59 km2 or 259 ha), and they will decrease slightly toward the north. Officially all regular sections cover 640 acres in the north as well as in the south.

The towns (in Wisconsin the name for civil townships) were established at different periods and some of them transferred sections to others. At present only the towns of Belmont, Lanark, New Hope and Pine Grove correspond to congressional townships with 36 sections. Alban and

Almond, which also corresponded to congressional townships, lost some of their land when the villages of Rosholt and Alban were incorporated.

In conformity with the U.S. Land Survey, Portage County with its 22 townships each measuring 36 square miles (94 km2) should have an area of 792 miles square (2,050 km2) or 506,880 acres (205,000 ha), but these figures are too high due to the curvature of the globe. However, the figure given by the Department of Natural Resources, 522,137 acres (211,301 ha), and the figure given by the U.S. census, 810 miles square (2,097 km2), are both higher. This is explained by errors present in the first surveys. In 1839, Joshua Hathaway surveyed part of Wisconsin but only a few miles of what is now Portage County. The complete and official survey of Portage County was made between 1851 and 1853 by a number of men contracted by the Surveyor General's Office. They were: Caleb W. Booth, Ira Cross, John E. Davidson, John D. Evans, W.P. Huntington, James M. Marsh and William J. Neely. These men took necessary measures in the field and reported their findings to the Surveyor General's Office in Dubuque. There, maps of each township were established at a scale of 10 chains to an inch. These maps were on forms entitled: Township No --------- Range No -------- Meridian ------ which for instance when filled would read: Townships No 23 North Range No10 East of the 4th Principal Meridian Wisconsin. At the bottom of the map was printed the following:

Total Number of Acres:

Surveys Designated-By Whom Surveyed-Date of Contract-Amount of Surveys- When Surveyed

M. Ch.s Lk.s

Township Lines

Subdivisions

At the right of the map was printed:

The above Map of Township No ------------ Range No ------------------ Meridian ----------- is strictly conformable to the field notes of the survey thereof on file in this office which have been examined and approved.

Surveyor General's Office ________________

Surr Genl

____

When filled, the form for one map might state: Total Number of Acres: 22,905.51 (9,276 ha)
Surveys Designated-By Whom Surveyed-Date of Contract-Amount of Surveys-When Surveyed
M. Ch.s Lk.s

Township Lines-Ira Cross-May 9"1851"-23"-78"-51- June 1851
Subdivisions-John D. Evans-Nov 15"1851"-59"-75"-26-Nov&Dec 1851

The above Map of Township No 23 North of Range No 10 East 4th Principal Meridian Wisconsin is strictly conformable to the field notes of the survey thereof on file in this office which have been examined and approved.

Surveyor's General's Office signed Geo. B. Sargent
Dubuque April 10 1852 Surr Genl

When these early surveyors did their measurements, they worked in an area largely wooded with many streams and lakes. The land was almost uninhabited, roads and trails were scarce. Tools were not as sophisticated as modern instruments, and it is also possible that some surveyors' work was not accurate. The survey of Portage County was started both from the east and from the west. Normally with the township lines at one mile intervals, surveyors from each side would meet, and the range lines being parallels would also join. But this was not the case, the township lines from the east were arriving about a quarter mile (half a kilometer) south of the lines traced from the west. And the townships measured from the east arrived at about one mile (1.6 km) from townships measured from the west. In addition, the township lines measured from the south left a gap between the townships and the Marathon County line. There was no question that a new and accurate survey must be taken. A compromise was subsequently adopted by which the township line in Range 9 East, instead of being exactly east-west would be slightly south-east to north-west; and that in the same Range 9, East sections 6, 7, 18, 19, 30, 31 would be almost double those in Ranges 9 and 10, Township 25--thus the area of sections would be larger than 640 acres (255 ha). This was accepted by the Surveyor General and on each map of townships located in Range 9 East, the paragraphs dealing with the surveyors' contracts, with the wording "the map is strictly conformable to the field notes of the survey...", were duly signed by the Surveyor General.

At that time (1851-1852), most of the land in Wisconsin was owned by the U.S. Government. Civilians could either buy tracts of land from the government through land offices (such a land office was established in Stevens Point in 1853) or they could homestead--in which case they had to reside upon and cultivate their piece of land six months of the year. At the expiration of five years, homesteaders had to prove they still lived on their piece of land. Land was also granted to certain officers and soldiers who had been engaged in United States military service. A number of corporations bought tracts of land from the government and then sold forties to immigrants. For instance, the tax assessment roll of the Town of New Hope shows that the "Fox and Wisconsin Improvement Company", a company that was to develop the Fox and Wisconsin Rivers as a means of opening up further communications with the interior, in 1866 owned 9,806.66 acres (3,968 ha) in what is now the Town of Alban and many lots in other towns. As the original survey was not accurate, the so-called forties varied. For that reason when describing a piece of land to be transferred between buyer and seller, most indentures close with the words: "according to government survey, more or less." The areas have been surveyed again, and although the limits of the sections and subdivisions remain the same, the areas are now given their true value in the modern plats and the tax assessments are based on these figures.

According to the U.S. Census, the area of the county is 810 square miles (2,100 km2) which is equal to 518,400 acres. This figure is somewhat lower than the data given by the Department of Natural Resources, but no explanation is given for the discrepancy.

GEOLOGY

As with the geology of any region, it is necessary to go back to the origin of the continents to understand fully the geology of Portage County.

Scientific research in the field of geophysics advanced rapidly during the 20th century, and today we have coherent theories that explain how the continents reached their present stage. We shall attempt to summarize them here.

When the earth was formed, it was a mass of hot liquid matter--magma enveloped by vapor. The external part of this mass became solid; the earth's crust formed into two layers of igneous rock. The upper layer was of minerals, relatively light in color and in weight, consisting mostly of silicon and aluminum, and called sial. The lower layer, darker and heavier, consisting also of silicon but with other dark minerals, mostly magnesium and iron, was called sima.

Part of the vapor enveloping the globe became liquid and formed an unlimited ocean. The rest of the vapor remained gaseous and became the atmosphere. One part of the earth's crust emerged as a single landmass, named Pangaea meaning "all the land". This was located in the southern hemisphere. As soon as the land emerged, weather elements (temperature, wind, precipitation) acted on it and thus began the processes of weathering and erosion.

Streams carried material resulting from weathering (boulders, pebbles, gravel, sand, silt) into the ocean where they were deposited, and under pressure of water, became rocks. Tectonic forces acted on these rocks and pushed them into domes that emerged from the water. These new formations were in turn subject to folding and faulting; forming ranges, mountain blocks and valleys.

The earth is still the object of such a cycle: land, and especially mountains, weathering; disintegrated material being carried by streams into bodies of water; land emerging; mountains rising. In addition, consider new mountains formed by volcanic activity or by faulting; new depressions resulting in subsidence or faulting, some of which filled with water and became seas or lakes.

Sial because it is lighter than sima, is not stable. At an early time, Pangaea broke up into two landmasses; one called Gondwana that remained near the South Pole in the Southern Hemisphere; the other call Laurasia which moved northwards. Geologists studied the various layers of rocks and in the 17th century divided them into eras based on the type of faunas that were attested to by fossils in the various layers of rocks.

The Primary era is traceable to fossils of invertebrates, fishes and amphibians, indicating the first traces of life. Fossils of reptiles, birds, dinosaurs appeared in the Secondary era; fossils of mammals were found in the Tertiary era. Finally the Quaternary is characterized by the appearance of man.

Each era was divided into periods, also called systems, according to their characteristics. For instance, the Primary era comprised the following periods: Cambrian, Ordovician, Silurian, Devonian, Carboniferous (subdivided into Mississippian and Pennsylvanian), and Permian. In modern times, traces of life were found in rocks older than the Cambrian. These were called Pre-Cambrian. Eras had their names changed into Paleozoic (ancient life) for the Primary; Mesozoic (intermediate life) for the Secondary; Cenozoic (recent life for the Tertiary and Quaternary). However, Tertiary and Quaternary remained unchanged as names of periods.

Periods are divided into epochs or formations (epochs referring to major subdivisions, and formations to groups of related strata that were formed in a geological period). In Wisconsin, the recognized formations of the Cambrian are the following: Mt. Simon, Eau Claire, Dresbach, Franconia, Mazomanie, Trempealeau and Madison. The Quaternary Period is divided into the Pleistocene and the Recent periods. Geologists have roughly dated eras, periods and epochs. According to them, the solid earth crust started about four billion years ago; the Pre-Cambrian age dates back to nearly three billion years ago to about 500 million years ago when the Paleozoic era started. The Cambrian period started 500 million years ago and lasted until 410 million; the Paleozoic era ended 200 million years ago when the Mesozoic era started. This era lasted until 70 million years ago, which marks the beginning of the Cenozoic era. This had one epoch, the Pleistocene which lasted from one million years ago to 10,000 years ago. We are now in the Recent epoch.

The earth's crust is formed of a dozen plates--a few large ones and several small plates which are fitted against each other along boundaries of three kinds: ocean ridges along which plates move apart and new crust is generated; transformed faults which allow plates to slide past each other; and subduction zones where plates collide with the result that the edge of one plate dives under the ledge of the other and crust is destroyed. Most earthquakes occur in belts which lie along the tectonic plate boundaries. This is also where most volcanic chains are found.

It is at the end of the Paleozoic era and the end of the Permian period that Pangaea divided into two continents: Gondwana comprising Antarctica and land that was later detached from it, and Laurasia, a large mass comprising all the rest.

Laurasia consisted of a number of plates. As it drifted northward it divided; continents broke apart and drifted away. Evidence of these movements, called the Continental Drift, is clearly visible in the coastline of eastern South America and western Africa (or North America and Europe) which would fit together as pieces of a puzzle. But other evidence is found in studying the similarity of rock formations in South America and South Africa; and the study of paleontology, paleoclimates and paleomagnetism.

Laurasia was made up of some large domes such as the Laurentian shield, the Fenno-Scandian shield, and the Ankara shield. The subsidence of land had formed seas or geosynclines in which the deposition of material formed the strata of the rocks of the Paleozoic era.

Portage County, as Wisconsin, was part of the first emerged landmass. Over what is now North America, a dome formed; it was the Laurentian shield of crystalline rock, granite that was in part transformed into a metamorphic rock called gneiss. Later during the Cambrian period, a shallow sea covered the area of Portage County. The material deposited in the sea during that period, and that eventually formed the strata of Dresbach, indicate a marine environment near the shore. Ripple marks and crossbedding in superposed layers, as can be seen in the quarry near Ellis, demonstrate that there were several stages during which currents changed. Before the end of the Cambrian period, the land emerged in only one formation. It may have been caused by the rising of a dome in northern Wisconsin by which all strata were lifted.

In Portage County, the bedrock slopes gently downward from the northwest to the southeast at a rate of about ten feet per mile (5 m. per km). As soon as the land emerged, the onslaught of erosion first created ridges and valleys, then slowly flattened elevations and filled depressions to form a peneplain. Isolated monadnocks such as Rib Mountain, the Mosinee Hills and Hardwood Hill in Marathon County, resisted erosion because they were capped with hard quartzite.

The uplift of the Wisconsin dome resulted in the formation of cuestas: the Prairie du Chien cuesta which is the first, followed by the Niagara cuesta and later by others. No cuesta is found in Portage County, but there are a few outliers of the Prairie du Chien cuesta (formerly called Magnesian cuesta) which extend in a curved shape from Marinette County southwest to Dane County and then west to Prairie du Chien. The closest part of the escarpment is about 35 miles (56 km) west near Winchester but three outliers stand in Portage County as isolated hills. Mosquito Bluff, near Almond is the most conspicuous. About a half mile (.4 km.) northwest of Mosquito Bluff is The Ledge. The third one in the county is Lone Rock, three and a half miles (5.6 km.) southwest of Mosquito Bluff near the Waushara County line.

It is likely that before the glaciation there were more outliers but they were destroyed by the advancing masses of ice. The quarry near Ellis, where sandstone is present near the surface of the outwash plain, was probably one of them. The Prairie du Chien cuesta had its origin in a stratum of dolomitic limestone which laid above the sandstone and acted as a caprock resisting weathering. However, the outliers in Portage County have been stripped of caprock and consist only of sandstone.

With the exception of the outliers, the bedrock in the county represents the old peneplain. As the strata are slightly inclined, the north part is of Pre-Cambrian metamorphic and crystalline rocks: granite, diorite, gneiss--and especially in the northeast, basaltic to rhyolitic metavolcanic rocks. In the central and in the southern part, there is Cambrian sandstone, but only sandstone of the lowest strata--the Dresbach formation. It is only in the northwestern part of the county that the bedrock is close to the surface and a few outcrops are visible. Consisting of hard rock, they are the remnants of former ridges and form the rapids of the Wisconsin River between Stevens Point and Plover and the subdued hills in the town of Linwood. The ages of these rocks have been dated and were found to be the oldest known in Wisconsin, originating approximately one billion 940 million years ago.

We have no evidence of what happened to central Wisconsin between the end of the Cambrian period, 410 million years ago, and the beginning of the Quaternary or Pleistocene, one million years ago. It is not without possibility though, that some day excavation under the drift may uncover bones of extinct animals: dinosaur, archaeopterix, pterodactyl; or pollen of primitive plants or trees which may have been here during the Mesozoic era and the Tertiary period.

During the Pleistocene, a large part of North America and Europe and part of northern Asia were covered by enormous ice sheets. Four major advances and retreats (called glaciations) occurred, spaced over a total period of about one million years. These four glaciations have been identified by the traces of round forms of deposits that the glaciers carried and deposited as moraines. In America, moraines were named after the regions where they were most conspicuous: the first one, the Nebraskan where the moraines were concealed under the deposits of the second, the Kansan; the third was the Illinoian, and the last the Wisconsin. Though there is no clear evidence, it is likely that the first three also covered Wisconsin.

The Wisconsin glaciation had several stages of advance and retreat. It is the Woodfordian stage characterized by an enormous mass of ice (also known as Cary stage) that largely gave landforms of Portage County their present shape. Because there was a dome in northern Wisconsin, the major ice flows went on either side. Little ice passed over the highland, the glacier did not reach the central part of the state, nor the Mississippi Valley from the confluence of the Chippewa River and the Mississippi River to the Illinois line. But lobes of the glaciers met again further south.

The eastern section of the Wisconsin glacier is the Green Bay lobe. Its end moraine--the terminal moraine of the Woodfordian stage also called the Outer moraine--forms hills running north-south, constituting the continental divide between the waters flowing to the Mississippi River and the Gulf of Mexico; and the waters flowing to the Great Lakes, the Saint Lawrence and the North Atlantic Ocean. From this range another line of hills emerges near Custer and runs southward. It is the Arnott moraine which is the terminal moraine of an older stage of glaciation. No evidence has been found to determine to which glaciation it belongs. Some believe that it goes back to the Illinoian; others estimate that it is the terminal moraine of an early advance of the Wisconsin glaciation. In the north and west of the county are traces of moraines of the Wisconsin glaciation, but no clear traces of terminal moraines. The central part of the county and its expansion southwest of Almond towards Waushara, Adams and Wood counties, is an outwash plain formed by material carried by streams formed from melting glacial ice. With the warming of the climate, the Woodfordian glacier melted and receded.

Retreats alternated with advances, marked by deposits of new end moraines which were identified as recessional moraines. The massive end terminal moraine of the Woodfordian glaciation blocked the Wisconsin River near the Baraboo Range and created Glacial Lake Wisconsin. The lake covered a large extension: unglaciated land, outwash plain and glacial drift. At times it probably extended to the southern part of Portage County, especially during floods. Later the lake shrunk, and vegetation grew on the lakeshore, eventually turning into peat.

As mentioned, the Woodfordian stage was characterized by thick and heavy masses of ice scraping the bedrock under it and transporting material over long distances. Chunks of copper originating in the area of the Michigan Upper Peninsula have been discovered among material deposited by the glacier in central Wisconsin. In Portage County and its vicinity, the glacier scraped the top of bedrock ridges, detaching large boulders and forming boulder trains still visible--especially in the towns of New Hope and Lanark. Some of the material carried by the glacier formed egg-shaped hills under the ice known as drumlins. There are only a few drumlins in Portage County--in Amherst and Lanark townships. They are relatively small compared to the large drumlins found in fields near Fond du Lac and between Madison and Milwaukee.

When the glaciers started to melt, some large blocks remained among the boulders and the other material. When these blocks melted they formed kettles, some of which eventually filled with water. The water coming from the melting ice carried much lighter material which was deposited in front of the glacier, forming the outwash plains. Streams, in irregular patterns, found their way across the drift. Only a few of these streams remain, such as the Plover, the Little Plover, the Hay Meadow Creek, the Tomorrow and the Flume. Others dried up but left lakes in strings occupying depressions in their former beds. Such are the lines formed by Stoltenburg Lake, Skunk Lake, Minister Lake, Sunset Lake and Hintz Lake--or by Rinehart Lake, Onland Lake, Reton Lake--or by Lime Lake, Adams Lake, Bear Lake--or by Pickerel Lake, Pine Lake, Pleasant Lake, Poplar Lake and Wolf Lake.

When the Woodfordian ice mass receded, vegetation began to grow on the deposited material, however, it was not the end of the great glaciations. Once more, the Wisconsin glaciation advanced--the Valders stage. It did not reach Portage County but remained in the east. Its outwash plain, however, reached the county. The glacial drift buried the vegetation. No trace of it was found in Portage County, but a forest covered by the glacier became petrified and was discovered near Two Rivers. Dated by the carbon method, the forest could be said to have existed nine and one half million years ago. With the retreat of the Valders ice stage, about nine million years ago, Wisconsin became totally free of ice. Today in America, Greenland marks the last land area covered by the continental glacier.

Differences in sea level indicate different epochs in the life of streams marked by terraces which are remnants of former flood plains. They are especially evident along the Wisconsin River, but also along other streams.

Eolian deposits, in the form of loess, are found in thin layers and in some sand dunes.

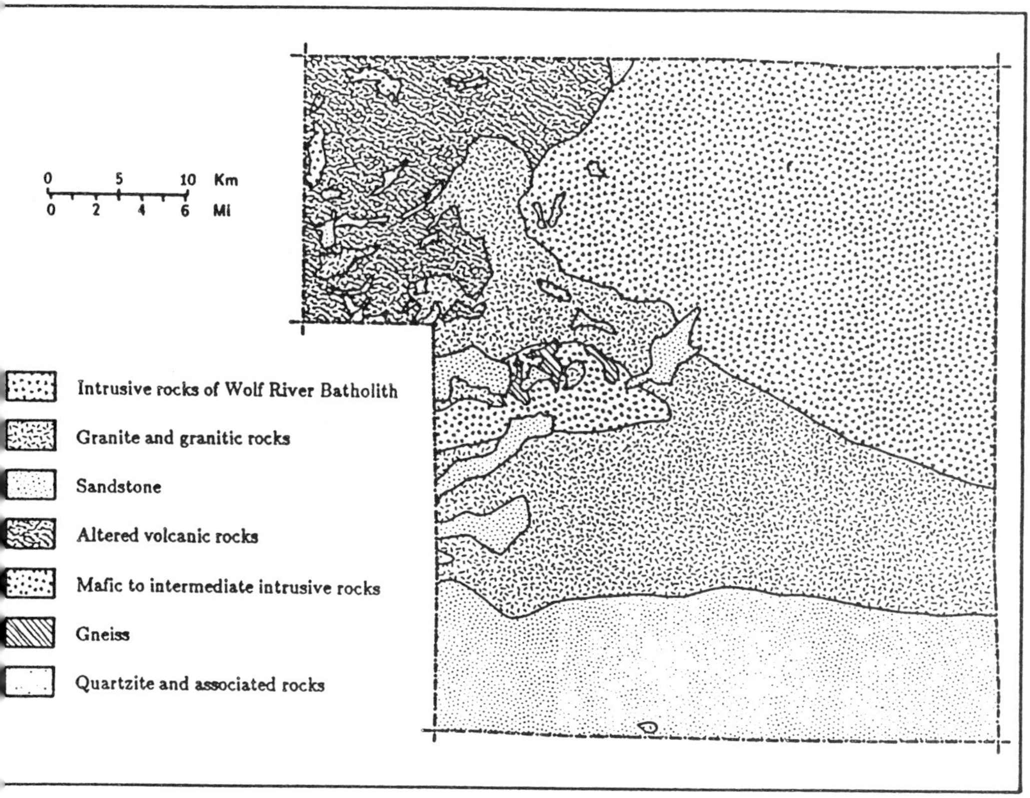

GEOLOGY: BEDROCK

Source: Geological and Natural History Survey

LANDFORMS

Portage County is divided between two major physiographic regions of the country: the Northern Upland and the Great Plains. The boundary is the limit between the area with underlying crystalline rock and the area with underlying sandstone.

One may, however, divide it into three regions:

1. The region of the old drift in the northwestern part.
2. The region of the young drift in the northern and eastern part.
3. The outwash plain and lake bed of former Lake Wisconsin in the central and southern part.

The region of the old drift:

Before glaciation, the area was not flat, but there were some low ridges and valleys. The drift has covered most of it. The surface now is almost flat but some of the old ridges, especially in the town of Linwood, appear as lines of upland above the low areas which are, in part, occupied by marshes and swamps. The Wisconsin River flows through the region. Its action has removed the old material and dug into the crystalline rock, which is visible south of Stevens Point where it forms rapids. It has also, in several places, formed floodplains which are about 20 feet (6 m) below the drift plains. One accumulation of old drift in the central part of the county is the Arnott moraine. However, we shall consider it as an isolated feature (as sand dunes and outliers), an enclave within the outwash region.

The region of young drift:

This region is delimited by the terminal moraine of the Woodfordian period of the Wisconsin glaciation. Presently, this moraine forms a series of hills, knobs and ridges. The longest and most conspicuous has a moderate slope on the outer side on the west. On the eastern smoother side it extends from the western edge of the town of Sharon to the eastern side of the town of Almond, passing between Ellis and Polonia, Stockton and Custer and Bancroft and Almond. In several places there are gaps where the material has been removed by the waters coming out of the glaciers. In the terminal moraine there are several low ridges representing recessional moraines and a number of kettles, some of which now contain small lakes.

Behind the terminal moraine the land was covered by drift, which explains the undulating topography, with hills and depressions being the material brought when moving ice was deposited irregularly. A few hills are drumlins, or streamlined hills consisting of glacial till with conspicuous steep and gentle slopes, the steep slope pointing the direction from which the ice came. There are few typical drumlins. The best examples are located in the town of Lanark, notably in Section 19. In the moraines, there are a number of kettles formed by melting ice blocks that broke away from the glacier and were deposited along with boulders, sand and gravel. Part of the ground moraine was removed by waters coming out of the receding glaciers. These waters deposited new material, forming outwash plains. In this region of young drift are pitted outwash plains--plains with holes which, like kettles, are the result of melting ice blocks. There is no clear division between the terminal moraine and the ground moraine.

The outwash plain and the former lake bed:

The central plain is underlain by sandstone bedrock of the Upper Cambrian age. This period comprises layers of different hardness. They had lain flat in a shallow sea, but when the dome of northern Wisconsin rose, these sediments, affected by the movement, rose towards the North. Erosion acted only on the hardest layers, causing partial erosion and forming cuestas.

As noted, there is no cuesta in Portage County, but as the front of the Prairie du Chien cuesta eroded a number of outliers remained. Three of them, all located in the town of Pine Grove, are found in Portage County. Mosquito Bluff or Mosquito Mound, made up of two hills about two miles (three km) south of Bancroft, is the most conspicuous. It is also the highest, having an elevation of about 1,245 feet (378 m) above sea level and about 150 feet (45 m) above the plain. The Ledge, about three-and-a-half miles southwest of Bancroft, is a crenelated rock about 1,180 feet (360 m) above sea level and about 100 feet (30 m) above the plain. Six hundred feet (180 m) to the northeast is a small hill only 20 feet (6 m) high which seems to have been part of the same outlier, the northernmost outlier in the county. One more outlier of the same Prairie du Chien cuesta is Lone Rock, three-and-a-half miles (4 km) southwest of Bancroft, with a top at about 1,150 feet (350 m) above sea level and 75 feet (23 m) above the plain. Both The Ledge and Lone Rock have northeastern to southeastern orientation whereas Mosquito Mound is oriented east-west and does not seem to be related to the other two. There were probably more outliers in the peneplain before glaciation, but the ice and strong winds eroded the surface rocks.

During glaciation, and for some time after, the ice retreated. Because the land was barren, the winds transporting sand strongly eroded the rock surfaces. The prevailing winds in Portage County come from the northwest and the west; it is the winds from the west that bring sand. Some eolian material coming from the Sand Hill region of Nebraska formed thin layers of loess, although some loess in the southern part of the county may have formed in the lake bed of glacial Lake Wisconsin. Sand transported by wind also comes from the alluvial plains of large streams and from the lake beds of former glacial lakes. When there was no vegetation, the sand was deposited in the form of dunes. The lifting power of the wind is very slight, and as sand is about 2,100 times heavier than dry air, it is moved in a sliding manner. The grains of sand, in contact with the land surface, diminish in size and some become dust that can be carried long distances. Consequently, the sand grains that form dunes or other accumulations have not travelled far.

In Portage County there are three dune fields. Two dune fields in the northern part are smaller in size, while the remaining field in the south is much larger. In addition to the above, there are several isolated dunes in the northern half of the county. One dune field is located along the east side of the Wisconsin River, about five miles (8 km) north of Stevens Point. It is a beltlike formation of sand ridges and dunes. The ridge ends in a large sand hill which is then dissected by a watercut gully north of it, resuming a ridgelike formation for another quarter of a mile (400 m). The dunes are about 20 to 30 feet (6 to 9 m) in height above the alluvial plain with the exception of the sand hill which rises up to 75 feet (23 m). In connection with the dunes, there are some blowouts where the sand has been removed by the wind to form depressions, some of which reached the crystalline bedrock under the sand cover. The larger dunes are transverse dunes; the smaller are mounds. The source of sand for all, is from the Wisconsin River floodplain.

North of Junction City, in the town of Eau Pleine, lies another dune field. These are all transverse dunes oriented in a west-northwest, east-southeast direction. The largest dune measures about 50 feet (15 m) by 1,660 feet (50 m), but most of the others are small, measuring about 10 feet by 500 feet (3m x 150 m). Most are under 10 feet (3 m) in height with a exceptions. All lie on an alluvium base or at the alluvium-crystalline base margin. From their orientation it is most likely that they were formed from sand from alluvium materials deposited there by streams resulting from the water erosion of the crystalline uplands and glacial outwash sands, and transported by south-southwest winds.

The largest area of sand dunes is south of Plover. There are two kinds of sand dune formations in this area. The first is large, concentrated in a line extending for two miles (3 km) from Coddington in a northeastward direction. It forms a ridge about 1/8 of a mile (200 m) wide and 2 1/4 miles (3.6 km) long, consisting of mounds about 13 to 18 feet in height (4 to 5.5 m). The highest mound, located immediately south of Coddington, is about 23 to 25 feet (7 to 7.5 m) high. Just north of this ridge of sand mounds are two long transvers dunes in an east-west direction, extending together for 1/2 a mile (800 m), about 10 feet (3 m) high and 60 feet (18 m) wide. They were probably constructed by strong south winds that blew sand from a dried-up lake bed of former Glacial Lake Wisconsin. A large area in the towns of Buena Vista and Grant, including Coddington, is an area of deflation where there are traces of numerous blowouts and associated rim-like dunes. The blowouts range in diameter is from very small (under 100 feet) to quite large (over 600 ft), but all are under 10 feet (3 m) in depth.

There are five isolated sand dunes in the county, four are transverse sand dunes, the last one is a ridge of sand mounds. Three are located on the western side of Mill Creek; the southern one lies partially on crystalline rock and partly on alluvium. The second is on crystalline rock with its northern end on alluvium. The third one is on alluvium. In the Dewey Marsh there is a long ridge of sand mounds orientated in a general southwest-northeast direction, as in the ridge near Coddington. It is about 300 to 350 feet (90 to 100 m) wide, 20 to 30 feet (6 to 9 m) high and about 1 1/2 miles (2.4 km) long. The whole line lies on a base of alluvium. South of this line lies the fourth transverse sand dune, which is about 10 feet (3 m) high and 50 feet (15 m) wide and orientated in a northwest-southeast direction. This transverse sand dune, as well as the sand mounds, were formed from sand of the alluvial plain blown by south-to-southwest winds.

Although Portage County has no mountains and seems to consist of a large plain, it has a number of hills and depressions: moraines, kettles, outliers of a cuesta, sand dunes and wind blowouts.

Geographers classify landforms according to various parameters, and those which apply here are (1) type of landforms: smooth plains, plains with hills, rolling land and hilly land; (2) origins: flood plain, outwash plain, end and recessional moraines and ground moraine; (3) type of slope--here mostly gentle; (4) local relief (difference between the highest point and the lowest point in a given area) and (5) type of cover.

We find in Portage County 12 classes of landforms:

1. Smooth plains-outwash plain, with 0-30 feet (0-10 m) of local relief, over 75% of gentle slope.
2. Smooth plains-flood plain, with 0-30 feet (0-10 m) of local relief, over 75% of gentle slope.
3. Plains with hills-outwash plain, 30-90 (10-30 m) feet of local relief, over 75% of gentle slope.
4. Plains with many hills-outwash plain, 30-90 (10-30 m) local relief, 25-75% of gentle slope.
5. Plains with hills-ground moraine, 30-90 feet (10-30 m) local relief, over 75% gentle slope.
6. Plains with many hills-ground moraine, 30-90 feet (10-30 m) of local relief, 25-75% of gentle slope.
7. Rolling land with hills-flood plain, 30-90 feet (10-30 m) of local relief over 75% of gentle slope.
8. Rolling land with hills-outwash plain, over 90 feet (30 m) of local relief, 75% of gentle slope.
9. Rolling land with hills-ground moraine, over 90 feet (30 m) of local relief, over 75% of gently slope.
10. Irregular plains-ground moraine, 0-30 feet (0-10 m) of local relief, 75% of gentle slope.
11. Hilly land, outwash plain, over 90 feet (30 m) of local relief, 25-75% of gentle slope.
12. Hilly land-end moraine, (terminal or recessional) over 90 feet (30 m) of local relief, 25-75% of gentle slope.

In the region of the old drift, there is smooth plain along the Wisconsin River in the Little Eau Pleine marsh. The rest consists of plains with hills and irregular plains. In the region of the young drift there are plains with many hills (either on the ground moraine or outwash plain) in the northern part, and mostly rolling land (either on the ground moraine or outwash plain) in the southern part, but most of the region is hilly due to the end moraines. The outwash plain region is a smooth plain on which rise the few outliers of the Prairie du Chien cuesta.

HYDROGRAPHY

Before glaciation, drainage was regular or dendritic, i.e. a treelike pattern where one main stream is fed by a number of tributary streams, which in turn are fed by smaller tributaries. With glaciation came ice, carrying rocks and sand to fill the stream valleys. Once glaciers retreated, the remaining material formed moraines and outwash plains. An irregular drainage pattern was forged to make up the Wisconsin River, smaller streams, rivers and creeks, swamps and marshes and numerous lakes.

A distinction can be made between surface water and underground water. Both are divided into two basins: the Wisconsin River Basin, the most important which drains the western two-thirds of the county and the Waupaca-Little Wolf River Basin which drains the eastern third. The boundaries of these basins show some differences between the groundwater and the surface water. The groundwater divide between the Wisconsin and the Waupaca rivers follows the general trend of the outer moraine, whereas the surface water is divided by the second moraine. The groundwater divide is marked by an area of hills and an undrained depression covering about 250 square miles (650 km2) extending two to three miles (3-5 km) on both sides of the divide.

With the exception of the Wisconsin River and the Plover River, most of the water in the county comes from precipitation (rain, snow, sleet, hail). Much of the precipitation is then returned to the atmosphere through evaporation from soil and transpiration from plants. The remainder, the runoff, flows overland to surface streams, replenishes soil moisture or percolates downward to the region of saturation; the surface of which is the water table.

Surface waters are the streams, the lakes and the wetlands. The Wisconsin River flows into the Mississippi River which flows into the Gulf of Mexico. The Little Wolf River and the Waupaca River are tributaries of the Wolf River; their waters flow to the Fox River, Lake Michigan and the Saint Lawrence River ending their course in the North Atlantic Ocean. The boundary between the Wisconsin Basin and the Waupaca Basin can therefore be considered a continental divide. Indeed, in northern Michigan and northern Wisconsin, markers have been erected indicating this continental divide.

In Portage County there is some difference between the surface waters and the groundwater divide because part of the divide is marked by flat areas partially covered with swamps, and a strip (including the village of

Almond) where the groundwater flows to the Waupaca Basin and the surface water flows toward the Wisconsin River. In some places the surface water watershed is clearly marked by the crest of the ridge formed by the terminal moraine of the Wisconsin glaciation. This is conspicuous by the slope at the plain's edge on the highways from Stevens Point to Polonia and to Custer.

The headwaters of the major stream, the Wisconsin River, originate in a network of interconnecting lakes and swamps in the northern highlands of Wisconsin near the Michigan border. Entering Portage County in the north from Marathon County, the Wisconsin River flows north-south to Plover, turns southwest and enters Wood County. On the right side, the Wisconsin River has several tributaries: Little Eau Pleine River, which starts in Marathon County; Rocky Run and Mill Creek, which starts in Wood County and receives Hayden Creek and Bear Creek. On the left of the river the tributaries are: the Little Eau Claire River, which starts in Marathon County; Hay Meadow Creek; the Plover River, which originates in Marathon County; the Little Plover River; Halladay Creek and Love Creek. In Wood County, the Wisconsin River receives streams draining the southwestern part of the county: Buena Vista Creek, which receives Four Mile Creek; Five Mile Creek; Seven Mile Creek and Ten Mile Creek.

The main stream draining the eastern part of the county is the Tomorrow River. It starts in the town of Sharon, in a swampy area at about 1170 feet (365 m) of elevation, then flows as a small creek across the moraine. Here the moraine is a landscape of knobs and kettles, some of which are filled with lakes. The Tomorrow enters the town of New Hope where it receives the waters of Poncho Creek, which passes mostly through swamps and marshes. The river enters Nelsonville Pond, formed by a beaver dam and later replaced by a man-made dam built to furnish power to a sawmill. The river moves on to enter the town of Amherst and Amherst Pond which also was formed when a dam was built to generate power for a mill.

The river's name, Tomorrow, is the English translation of the Indian word, Waubuck which has evolved into Waupaca; the name applied to the lower part of the river and to the county nearby. Generally, the name Tomorrow refers to the section of the river in Portage County, although the U.S. Geological Survey puts the name of Waupaca River to the section below Amherst. Here the stream receives the waters of Bear Creek, which is an emissary of Adams Lake and of Spring Creek, an outlet of Spring Lake. When the Tomorrow leaves Portage County it has an average flow of 100 cubic feet per second.

In Waupaca County, the Waupaca River has some tributaries that bring water from Portage County: Allen Creek, a small spring-fed stream that has a moderate flow, but has been known to dry up completely at times; Emmons Creek, originating as an outlet of Fountain Lake; Murray and Pearl Creeks starting as resurgences of groundwater (although in dry periods, their upper parts often have no water at all). These small streams have short courses in Portage County. Together they discharge about 60 cubic feet (0.03 m3) of water per second out of the county.

The northeastern part of the county is the basin of the Little Wolf River. It originates in Marathon County and is the drainage from swamps in Marathon and Shawano counties. Just before entering Portage County the Little Wolf receives Klondike Creek, emissary of Tree Lake. The river returns to Marathon County, comes back to Portage County and leaves it soon to enter Waupaca County. Its whole course in Portage County is only about one and a half miles long (0.8 km).

In Waupaca County, the Little Wolf receives streams from Portage County: Bradley Creek, the emissary of Lions Lake and Flume Creek, which originates in Marathon County and flows through Portage County for 15.6 miles (25 km) to form Rosholt Pond; Nace Creek, outlet of Rinehart Lake and Johnson Lake and Sannes Creek, outlet of Thorn Lake. The discharge of water from Portage County through the Little Wolf River amounts only to 60 cubic feet (1.68 m3) per second, about the same amount as the Waupaca River and its tributaries. The Little Wolf River flows into the Wolf River (which has received the Waupaca River and ends in Lake Winnebago). The waters of this lake reach Green Bay on Lake Michigan, from there to the Saint Lawrence and then, via Canada, to empty into the North Atlantic Ocean.

In addition to the streams previously mentioned, there are a number of short streams, some of which have intermittent flow. Streams may appear after heavy rainfall or melting snow carries their water to another stream, a lake or simply a depression in the ground. A temporary pond may be created, only to dry up when the weather warms.

Since the difference between the highest point, 1300 feet (209 m) and the lowest point, 918 feet (280 m), is small, the streams (with the exception of the Wisconsin River) are not significant, nor do they have strong currents. Only nine are more than ten miles (16 km) long. There are 35 named streams marked on maps of the U.S. Geological Survey.

It is not surprising, in a region that until about 10 thousand years ago was partly covered with glaciers which left large areas covered with drift, that there are many lakes in the county. According to a survey taken in the early 1970s, there are 135 lakes having a surface of more than 0.5 acre (0.40 ha). Since the survey was taken after a few years of relatively low precipitation, the number of lakes has understandably increased after a number of years with high precipitation.

The Wisconsin River, coming from a distance, receiving several tributaries and flowing in the vast Mississippi Plain, is considered to be in the maturity stage. It does not follow a straight course, but meanders across the lowlands. Being in a humid and temperate climate, the Wisconsin has high waters, especially in the spring when the snow melts, flooding the lowland and submerging the meanders. When the flood recedes, the stream may alter its course creating islands, sandbanks and lakes. Other streams, especially Mill Creek and the Plover River, also present meanders, meander cuts, islands and small floodplains.

A Wisconsin Department of Natural Resources report, published in 1972 and probably based on field studies of 1970 and 1971, lists 135 lakes in the county with an area of more than half an acre (2,000 sq m). The total area of the lakes is 4,015 acres (162.5 ha). As the measurements were taken after a number of rather dry years, the number of lakes over half an acre (2,000 sq m) and the total surface covered by lakes have increased, but in the future may decrease, with some lakes drying up completely to be covered with brush or grass. On an 1876 map there are a number of lakes, especially in the vicinity of Sunset Lake that no longer exist.

The 1972 Wisconsin Department of Natural Resources is used for the following statistics:

Lakes are of three origins:

- Natural lakes, by far the largest number, there are 96.
- Impoundments - lakes formed by a dam built by beavers or man. There are seven.
- Man-made lakes; there are now 22.

Natural lakes have developed in three ways: either in pits or kettles, which formed when blocks of ice mixed with boulders, gravel and sand, and carried by the glaciers, melted. These lakes generally have steep slopes. A second type was formed by the variations in the deposition of material carried by the glacier and causing depressions in the land, these have smoother slopes. The third type was formed by obstructions in the course of glacier-fed streams which created small basins or different channels; these were the origin of the chain of lakes. Although in Portage County there are no interconnected chains of lakes, there are concentrations of lakes which follow the former course of a stream. For instance, in line are Skunk Lake, Minister Lake, Sunset Lake and Hintz Lake. Further south are Lake Emily, Lake Julia, Ell Lake and Lake Thomas, and still further south are Pickerel Lake, Pine Lake, Pleasant Lake, Poplar Lake, Wolf Lake and Patterson Lake. They follow a branch in line to meet another branch with Kim Lake, Boelter Lake, Riley Lake, Silver Lake and Tamarack Lake. Some lakes have combined characteristics. Bear Lake is partly shallow but has one deep area that may have been a kettle. In Collins Lake there may be three different kettles.

These lakes have no drainage and represent the level of the water table, with their levels fluctuating at the same rate as the water table. A few lakes are fed by springs or small streams: Tree Lake, Lions Lake, Spring Lake, Adams Lake and some smaller unnamed ones. The enlarged drainage basins are the lakes that have both a tributary and an emissary, as: Lake Adams, Anderson Lake, Lake Helen, Lions Lake, Rinehart Lake and Tree Lake.

Oxbow lakes and bayou lakes are remnants of former meanders and channels of streams. They are found close to the Wisconsin River. In the spring when the snow melts, streams flood low areas along their shores and deposit sand and gravel that partially fill the original bed so that the stream may change its course. Consequently, there are abandoned channels and meanders which, in time, fill with earth and are covered with vegetation. Several oxbow and bayou lakes were submerged when dams were built on the streams.

Impounded lakes are created either through the work of beavers or men. The largest are on the Wisconsin River: the impoundment forming Lake DuBay, which is primarily in Marathon County; one in Stevens Point; one in Whiting and another in Plover. Another dam downstream in Wood County forms the Biron Flowage which extends into Portage County.

There are dams on other streams: one on the Eau Pleine River forming the Eau Pleine Flowage; four on the Plover River forming Dana Pond, Van Order Pond, Jordan Pond and McDill Pond; one on the Little Plover River forming Springville Pond; two on the Tomorrow River forming the Amherst Mill Ponds; the Nelsonville Pond (as the Nelsonville mill is no longer active, the dam has been eliminated and the pond has disappeared) and one on Flume Creek forming Rosholt Mill Pond. Impoundments existed on Flume Creek when sawmills were there and probably several were at one time on Mill Creek to accommodate sawmills.

Lake DuBay, the largest impounded lake, was formed when a dam was built in 1945. The lake covers approximately 6700 acres (2700 ha), with 40% of it in Portage County. The Wisconsin River Flowage, formed by the dam in Stevens Point, has an area of 2093 acres (85 ha). This flowage and the one in Whiting are the deepest lakes (maximum 25 feet - 7.5 m). Fountain Lake represents a special case, although it is a natural lake fed by springs with an outlet, Emmons Creek, a dam has artificially elevated the lake five feet higher than its natural level.

A number of lakes, 15 of significant size, were created when land was excavated for the construction of road ramps. Most are located along Highway 51. Geometrically shaped, the lakes are not deep, their depth merely corresponds to the watertable. The largest are Lake Pacawa with 36 acres (15 ha) and a depth of 12 feet (3.6 m); Lake Susan with 22.6 acres (9 ha) and a maximum depth of 23 feet (7 m) and Clear Lake with 17 acres (6.8 ha) and a maximum depth of 15 feet (4.8 m). Other small lakes have been manually excavated to enhance an area where houses were built.

All natural or excavated lakes are small. In 1971, the only lake covering over 100 acres (40 ha) was Lake Emily with 107 acres (43 ha). Only five lakes have between 100 and 50 acres (40 and 20 ha): Tree Lake with 71 acres (29 ha); Sunset Lake with 61.2 acres (25 ha); Bass Lake with 55 acres (22 ha); South Twin Lake with 52 acres (21 ha) and Lake Hintz with 51 acres (20 ha). The maximum depths recorded are in Lake Budsberg, 58 feet (17.6 m); Lake Collins, 56 feet (17 m) and Lake Sunset, 55 feet (16.7 m). East Lake was only two feet (30 cm) deep and Boelter Lake, five feet (1.5 m) deep.

The levels of lakes vary. When vegetation invades a lake, the water area decreases. The planimetric maps of Portage County, published by the U.S. Geological Survey between 1955 and 1958, indicate Patterson Lake, Washburn Lake and Tamarack Lake as sizeable surfaces of water. When the U.S. Geological Survey published topographical maps between 1968 and 1971, after a number of years of relatively low precipitation, Lake Washburn was identified as a marsh; Patterson Lake as a pond and Tamarack Lake was reduced to two small pools. In the early 1960s, Lake Patterson appeared to be completely dry and replaced by brush, but in the forest next to the former lakeshore was an abandoned boat--evidence that the lake once held enough water to provide fishing. After high precipitation in the late '60s and early '70s, the lake filled again and a photo showed water and a boat on shore. Fence posts under water in the middle of Lake Hintz indicate that at one time the lake bottom was pasture.

Wetlands are understood to be areas where water is at, near or above the land surface for long periods of time. In Portage County large areas can be identified as wetlands.

Wetlands are of many variations: swamps are forested areas; marshes are open areas and meadows are flooded only in the spring. Bogs form where decaying organic material has been deposited and a slough is a place of deep mud. Many bogs are remnants of shallow lakes filled with vegetation, or are at the edge of such lakes.

In reference to hydrologic formation, wetlands in Portage County fall into two types: surface water depression wetlands in which the watertable is below the level of saturated swamp or marsh (sometimes called a perched wetland), or groundwater depression wetland where the surface of wetland corresponds to the surface of the watertable. Many areas of the outwash plain or of the lake bottom of glacial Lake Wisconsin have become huge wetlands. Such are Buena Vista Marsh and Dewey Marsh. In the early part of the 20th century significant losses of wetlands occurred, especially in the southwest of the county when ditches were dug to produce agricultural land.

Wetlands are important to wildlife. They provide food, cover and nesting areas for waterfowl and shorebirds; habitat for furbearers and spawning ground for fish. Wetlands benefit the county since they reduce runoff and soil erosion; control flood, and improve water quality.

With the use of black/white and infrared aerial photographs, wetlands can now be inventoried. The Department of Natural Resources watches over wetlands use and directs their conservation.

The watertable varies according to the seasons. It is highest in the spring after snow has melted, but then decreases when vegetation takes water for growth. It is lowest in winter when the ground is frozen and does not allow water seepage. The watertable varies with precipitation and drought. However, only large amounts of precipitation causes a significant rise in the watertable levels or interrupts the seasonal decline. Water levels often rise in the autumn, when precipitation occurs after frost has killed vegetation. The watertable is high during long periods of above-average precipitation, and low during long periods of below-average precipitation. The levels were generally high in 1945-48, 1951-54 and 1960-61. They were low in 1949-51 and 1956-59. Levels do not fluctuate uniformly; they may rise in some parts of the county while they decline in other parts.

The watertable generally follows the trend of the land surface. On a map (Holt, plate 2) of the watertable based on wells, irrigation pits and staff gauges on lakes, two valleys following the trends of the Wisconsin River and the Tomorrow-Waupaca River can be seen. The highest point of the valley is under Lake DuBay at the Marathon County line, at approximately 1100 feet (335 m), and in the vicinity of the old Liberty Bell School at a level of 1105 feet (337 m). The lowest point is at the Wood County line under the Biron Flowage at approximately 1035 feet (316 m). A well in the vicinity indicates a level of 1039 feet (318 m). However, under some tributaries of the Wisconsin River in Wood County, the level is lower, at 1023 feet (312 m). It is approximately the same at Kellner under Four Mile Creek and Buena Vista Creek, and under Ten Mile Creek at the Wood County line.

Another valley is under the Tomorrow River; with its highest point in swamps near Twin Lake at approximately 1180 feet (360 m) and its lowest point at the Waupaca County line at approximately 965 feet (294 m). Wells in the vicinity measure a depth of 976 feet (297 m) and 978 feet (298 m).

Between the two valleys is a low ridge, with the highest point at the Marathon County line at approximately 1191 feet (363 m). Sloping down in a southern direction near Arnott, the ridge is at 1096 feet (334 m), rising again to 1140 feet (348 m), 1135 (346 m) near Saint Martin Church. The land declines again to 1112 feet (340 m), with a depression as low as 1099 feet (334 m) under Washburn and Bass lakes.

On the west side of the Wisconsin River, at the well near the old Runkle School and probably along a ridge to the Wood County line west of Junction City, the watertable reaches 1180 feet (360 m). The watertable forms valleys under other streams, noticeably under the Plover River, Flume Creek, Bear Creek (tributary of the Waupaca River) and Spring Creek.

The amount of groundwater in any area is determined by the geology of the area. Some rocks (aquifers) store water as reservoirs, and thereby supply water to wells and discharge water to streams, marshes and lakes.

Other rocks are impervious, and constitute the bottom and sides of the reservoirs.

Portage County is underlaid with crystalline rock of Precambrian Age and sandstone of Cambrian Age, which generally are mantled by glacial deposits of Pleistocene Age. In the northwestern part of the county, crystalline rock is exposed and weathered. It contains enlarged joints and fractures. Fractures may extend for some distance laterally, but they are generally less than 50 feet (15 m) deep. Crystalline rock is impermeable, but groundwater accumulates in fractures and shallow weathered zones. The surface of crystalline rock resembles a broad plain that slopes approximately ten feet per mile (1.8 m per km) from the northwestern part of the county to the south and southeast. The shape of the surface is that of a broad shield which slopes to the south. Sandstone, an aquifer, overlies crystalline rock in the southern part of the county. It decreases in thickness from an estimated 200 feet (61 m) along the southern boundary of the county, to zero feet in most of the central and northern parts of the county. Thick deposits of drift cover the sandstone except in areas where sandstone ridges or mounds protrude above the drift.

Other examples of aquifers lie in glacial drift, outwash deposits and postglacial deposits such as alluvium and sand dunes.

Areas where the character of rocks, the availability and quality of water, the topography and surface drainage are similar, are called groundwater provinces. Portage County can be divided into three such provinces: the sand plain province, the drift province and the crystalline rock province. The drift crystalline rock province consists of groundwater in thin aquifers of outwash, alluvium and fractured crystalline rock. The aquifer is thin, not more than 50 feet (15 m) thick, except in the fractures. The drift province consists of thick aquifers of sandy till, outwash, and ice-contact deposits. The thickness of the deposits ranges from a few feet to about 350 feet (106 m). The sand plain province consists of a thick and extensive aquifer of outwash sand and gravel. The thickness ranges from a few feet to about 250 feet (76 m) and averages about 100 feet (30 m).

In the spring, usually March and April, streams with high waters cover low-lying area. This can occur during other times of the year as well, in the event of heavy rainfalls. Occasionally, streams reach flood stage, i.e. water inundates areas where houses and roads have been built. The Wisconsin River has had a number of floods since the region was settled. The major floods occurred in 1847, 1866, 1880, 1881, 1905, 1911, 1935,

1938, 1941, 1960 and 1965, causing damages particularly in Stevens Point. Dams have reduced the risk of severe flooding and, with their use, the flow of streams can be controlled.

CLIMATE

Located halfway between the North Pole and the Equator, Portage County is in the temperate zone and in the midst of constantly battling air masses of polar and tropical origin. The polar air masses dominate during the winter months and the tropical air masses have their greatest influence during the summer.

Since Portage County is at a great distance from both the Atlantic and the Pacific oceans, it has a continental climate little influenced by those large water masses which cool in summer and warm in winter. Winters may be very cold and summers, hot.

According to the Koeppen classification of climates, Portage County has a Dfb climate, i.e., temperate and humid with severe winters; moist with no dry season (precipitation is distributed throughout all months); snow, with an average temperature for the coldest month being below - 3oC (26.6oF), and warm summers (more than four months with an average of 22oC (71.6oF).

Temperatures vary greatly from year to year. About once in three years temperatures will reach 100oF (37.8oC), while two winters in three will have temperatures of -25oF (-32oC) or colder.

The annual mean temperature for Stevens Point is 44.5oF (6.1oC). Coddington (1923-1960) averaged 42.4oF (5.7oC) in July 1936 and the lowest -38oF (-38.9oC) in February 1951.

In Stevens Point mean annual precipitation amounts to 31.4" (798 mm). Other parts of the county will not show any significant difference. Approximately 60% of the precipitation falls from May to September. The wettest month is June with 4.69" (118 mm); the driest is February with 1.26" (32 mm). There is an average of 115 days with one tenth of an inch (2.5 mm) of precipitation. There are about 20 to 50 days a year with thunderstorms, usually from noon to midnight with the most severe occurring between 2 p.m. and 7 p.m. Severe thunderstorms can develop into tornadoes. Although every year "tornado watches" are given in Portage County, there have been very few tornadoes--fewer than in the corridor of Wausau-Chippewa Falls-Hudson or areas in southern Wisconsin. Only one severe tornado has been recordedin the county. On Friday, August 20, 1863, a tornado destroyed the village of Buena Vista killing three people and leveling the post office, a tavern and a residence.

Annual snowfall averages about 48" (1220 mm), but seasons are highly variable with totals ranging from 102" (25 mm) to 90" (228 mm). The

first snow may fall in October and the last in May, but the average date for the first snowfall of one inch (25 mm) or more is November 15. The chance of one inch or more snowfall by October 20 is 10% and by December 10 is 90%. The probability of snow remaining on the ground increases until the middle of February and then decreases rapidly--although March may often be the snowiest month of the year. Snow melt accounts for about 6" (152 mm) of mean annual precipitation. Hail occurs only on an average of 2.5 days a year, but is usually small with little, if any, damage.

Predominent winds blow from the west and the northwest from late fall through mid-spring, and from the south for the rest of the year. Since winds are influenced by cyclones and fronts, there are winds from a variety of directions throughout the year: 30% of the time west through northwest, 30% south through southwest, 18% east through northeast, 18% south to southeast and the remaining 4% represents calm air. The strongest winds are westerly. Winds from 25 to 31 miles per hour (40-50 km) are present less than 1% of the time, 19-24 miles per hour (30-40 km) for 7%, 13-18 miles per hour (l0-30 km) for 23%, 8-12 miles per hour (13-20 km) for 30%, 4-7 miles per hour (6-12 km) for 22% and less than four miles per hour (6 km) 15% of the time. These are surface wind speeds reduced by friction.

At 30 feet (9 m) above ground, wind speeds can reach 60 miles per hour (l00 km). The windiest seasons are early spring and late fall, while the summer season is the least windy. Wind has an effect on the way cold is perceived by the body--this is referred to as "the chill factor." At a temperature of 32oF (0oC) one may feel cold, but with the same temperature and a wind of 5-15 miles (8-24 km) per hour, one will feel very cold--the sensation of cold is the same as if the actual temperature was 28oF to 15oF (-33oC to -26oC). With a wind of 15 to 45 miles per hour (40-70 km) one will feel bitterly cold, as if the actual temperature was 15oF to -8oF (-9.4o to -22.2oC). With a wind of more than 45 miles per hour (70 km) the exposed area of the body would be in danger of freezing. With a temperature of 0oF (-17.8oC) and no wind it would be cold, but with a little wind it would become bitterly cold. With a wind of more than 8 miles per hour (12.9 km), the exposed parts of the body would freeze in less than two minutes. In blizzards with strong wind gusts, the chill factor may be -100oF (-73.6oC).

In Portage County, all lakes and streams freeze in winter. Small, shallow ponds and lakes freeze first, then small streams with weak currents. The last to freeze are those streams with strong currents. The land freezes to a certain depth depending on the depth of snowcover. If no snow has fallen during a long cold period, the ground will freeze deeply. If there is snowcover, the ground will freeze slowly and at less depth.

The growing season in Portage County varies greatly from year to year, not only because of the weather, but also because of local topography, type of soil, type of vegetative cover and the land's proximity to cities, lakes or rivers. If 32oF (OoC) is considered "killing frost," the mean growing season is only 145 days in Stevens Point, 139 days in Rosholt and 100 days in Coddington. If 26oF (-3.3oC) is considered "killing frost," the growing season is 169 days in Stevens Point, 161 days in Rosholt, and 125 days in Coddington.

There are large areas in Portage County that have a potential for frost. Marsh lands with peat soils encourage the formation of frost. They are usually lower than the surrounding areas, are wet with little vegetation cover and tend to cool quickly at night because peat soil has a poor ability to store heat. Low areas may form basins in which cool air accumulates. Urban areas, however, discourage frost because of the amount of heat generated in that environment.

The mean annual humidity for Portage County is 70% to 80%. It is generally most humid at sunrise and least humid during the early afternoon.

During the year, Portage County receives from 2,400 to 2,600 hours of sunshine. From November to January "possible sunshine" is from 45% to 50%, from January to April it is from 50% to 60% and from May through October it is 60% or greater.

SOILS

The Chinese studied soils even before the time of Christ. In 1675, John Evelyn states that theorists of his day "reckon up no fewer than one hundred seventy nine millions one thousand and sixty different sorts of Earths."

But it is in the 19th century that the scientific study of soils began. The first to tackle the job were Russians who gave names to types of soils such as podzols, chernozems and sierozems which were used by soil scientists all over the world.

A new classification established by the U.S. Soil Conservation Service has been used in the United States since 1965. The hierarchy consists of ten levels. The number of classes recognized are: 10 orders, 47 suborders, 180 great groups, 960 subgroups, 4700 families and 11,000 series. In Portage County, 35 series have been identified.

Soil series are named for a town or other place name where the soil was first observed. Soils that are named after localities in Portage County are Kranski, Meehan, Oesterle and Rosholt soils. Altdorf, Coloma, Friendship, Mosinee, Plainfield, Rozelville and Vesper soils are named after towns in the vicinity of Portage County. Some series are further subdivided according to differences in texture or slope: Meehan loamy sand, 0% to 3% slope; Meehan loamy sand, sandstone substratum, 0% to 3% slope; Meehan sandy loam, red surface, 0% to 3% slope and Meehan fine sandy loam, gravelly variant. There are eight varieties of Rosholt soil.

Soil scientists, or pedologists, designate soil as surface material which is composed of both mineral and organic material. The underlying material or parent material is wholly mineral. For the agriculturist, soil is only the upper part in which plant roots may grow. Underneath is the subsoil.

Soil is formed through the disintegration of rock. The differences among soils relate to their formation and are primarily based on five factors: the parent material (rock which has disintegrated), the climate, the living organisms, the relief, and time.

Although Portage County appears uniform in physical features and in climate, it can be divided into three major regions: the old drift (some scientists consider this the driftless area), the younger drift and plain consisting of outwash plain, and the alluvial plain and bed of the former Glacial Lake Wisconsin. Each region presents a number of soil series differing in some of their characteristics. Some of these characteristics are: the time factor (old drift, new drift, alluvium of various periods), the

relief (although limited to a few areas), the parent material (bedrock, glacial drift or alluvial drift), the living organisms especially the vegetation: forest, swamp, marshes; and even the action of man (deforestation, agriculture, construction).

A cut through soil will reveal several layers, each with definite characteristics. Pedologists call these layers "horizons;" their characteristics form the base of the following classification.

Thirty-one series were identified in Portage County. Wet alluvial land, marshland, rockland are not considered in the study of soils. As the number of series and their variants is very large, a better view of soils in an area such as Portage County will be found by grouping soil series into soil associations. These form landscapes that have a distinctive pattern of soils. A landscape normally consists of one or more major soils and at lease one minor soil, and is named for the major soils. In Portage County, 11 such associations are distinguishable; they comprise four groups based on their formation:

A. Soils formed in loamy and silty materials and the underlying residuum from bedrock:
1. Kert-Norgo variant association
2. Neadland-Rozelville-Dolph association
3. Point-Dancy-Mosinee association

This group occupies the largest part of the old drift region in the northeast part of the county, on the west bank of the Wisconsin River and includes also the Arnold moraine.

B. Soils formed in sandy glacial drift:
1. Wyocena-Rosholt association
2. Kranski-Coloma-Mecan association

This group occupies most of the area which was covered by the terminal moraine of the last Wisconsin glaciation and large parts of what was the ground moraine behind the terminal moraine. Exceptions are areas where during the retreat of the glaciation, outwash plains developed above the glacial till or areas where sandy deposits of the glacial till existed.

C. Soils formed mainly in outwash sand and gravel, or sand:
1. Richford-Rosholt-Billett association

2. Plainfield-Friendship association
3. Leola-Pearl association
4. Roscommon-Meehan-Markey association

These associations almost completely cover the outwash plain and lake bed of former glacial Lake Wisconsin, and almost all the level areas that had been covered by the ground moraine of the Wisconsin glaciation.

D. Soils formed in alluvial or organic deposits:
1. Wet-Dunnville association and alluvial land
2. Markey-Seelieville-Cathro association

The first is found only in some areas along the Wisconsin River. The second also occupies some areas along the Wisconsin River and additional land in swamps near Bear Creek, the Plover River, North Creek and the Tomorrow River.

The soils represented in Portage County belong to 20 of the 47,000 soil families in the United States and to six of the ten large orders in which the soils of the world are divided. The only orders not represented here are soils found only in tropical or subtropical arid regions.

□

NATURAL VEGETATION

The presettlement vegetation cover has been identified from the field notes of Joshua Hathaway, who surveyed the so-called Indian Strip in 1839 and 1840. The strip was established following the Treaty of 1836 by which the Menominee Indians turned over to the Americans a territory three miles wide on either side of the Wisconsin River, and through the field notes of men who worked for the U.S. Land Survey from 1851 to 1853, which fixed the boundaries of the Territory of Wisconsin. In their field notes, surveyors described the trees they encountered at the corners of each section and each quarter of a section. They measured the tree's diameter at breast height, and recorded the direction and the distance of the nearest four trees. If no trees occurred within reasonable distance, pits and mounds of earth and sod were constructed. Every major change in vegetation, trees intersected by the survey line, rivers, streams, ponds, lakes, marshes, major topographic changes, windfalls, soil quality, and signs of human influence (or presence) were recorded in the survey notes.

On the basis of the field notes, four plant community groups, subdivided into 13 plant community types, can be counted. They comprise all the land cover of presettlement Portage County.

The first group is the Mixed Coniferous-Deciduous Forest which occupied 23 percent of the county, 182 square miles (473 km2) or 116,186 acres (47,262 ha). It comprises four subdivisions: 1) hemlock, sugar maple, yellow birch, white pine and red pine; 2) sugar maple, yellow birch, white pine and red pine; 3) white pine and red pine; 4) Jack pine scrub, oak forests and barrens (this subdivision occupied about ten percent of the county).

The second group is the Deciduous Forest. It comprises three subdivisions: 1) sugar maple, basswood, red oak, white oak and black oak; 2) white oak, black oak, bur oak and Hill's oak; 3) oak openings, bur oak, white oak, black oak, and Hill's oak (this group making up about 29 percent of the county).

The third group is Grassland and Brush. It represents only two percent of the county and comprises two subdivisions, prairie and brush.

The fourth group is the Wetland Vegetation occupying 27 percent of the county. It comprises three subdivisions: 1) swamp conifers (white cedar, black spruce, tamarack and hemlock); 2) lowland hardwoods (willow, soft maple, box elder, ash, elm, cottonwood and river birch; 3) marsh and sedge, meadow, wet prairie and lowland shrubs.

Major waterways and bodies of water occupied about one percent of the remainder of the area of the county.

Presettlement vegetation was the subject of the thesis David G. Tesch submitted for his Master of Science degree in the College of Natural Resources, University of Wisconsin-Stevens Point in May, 1982. Unfortunately, the map accompanying the thesis was not published. Rather, the map included in the book is an enlargement of G.E. Hanson's map of presettlement vegetation in Wisconsin which was reproduced in several reports since 1965. The literature cited lists 124 entries.

Pioneers who came to the area were loggers, followed by farmers who cleared land to produce crops for the loggers. The major assault on the forest took place after the Civil War. There was a great demand for timber to use as fuel, as well as a material to build cities, bridges, furniture and ties for railroads. At the turn of the century, as Wisconsin was leading the world in lumber production, the face of the forest changed dramatically.

Farming grew through four phases: providing food for home use and in logging camps, growing wheat, raising livestock and dairying, and the production of potatoes and crops for canning.

By 1939, lumber production had dropped to one-tenth of what it was in 1899. The land was cut over, and repeatedly burned over. Out of the Great Depression came federal work programs: the Civilian Conservation Corps (CCC) and Work Projects Administration (WPA). These work programs helped heal the scarred land and stimulated forestry work in the state. Today the people of Wisconsin are proud of their revitalized forests.

Most plantations of forests are less than 30 years old. Most softwood plantations grow: Jack pine, red pine, white pine, white spruce, black spruce, balsam fir, hemlock, tamarack, eastern red cedar and western white cedar. The hardwoods grown are: white oak, select red oak, other red oak, select hickory, other hickory, basswood, beech, yellow birch, hard maple, soft maple, elm, black ash, white and green ash, cottonwood, willow, bigtooth aspen, quaking aspen, paper birch, river birch, black cherry, black walnut and butternut.

In the county, the forested area covers 174,300 acres (70,417 ha). The Forest Service has divided the area into types named after the dominant species. Ten types are represented:

Oak-Hickory	48,000 acres (20,000 ha)
Aspen	35,100 acres (14,000 ha)
Maple-birch	32,200 acres (13,000 ha)
Elm-ash, soft maple	22,000 acres (8,800 ha)
Red pine	17,400 acres
Jack pine	4,400 acres (1,800 ha)
Northern white cedar	4,400 acres (1,800 ha)
Paper birch	4,300 acres (1,700 ha)
White pine	2,200 acres (890 ha)
Tamarack	2,100 acres (850 ha)
Non-stocked	2,200 acres (890 ha)

Almost all species of softwoods and hardwoods are found in the forests, but removal for commercial purposes is limited to a few species. From 1967 to 1982, the average annual removal was: 411,000 cubic feet (11,508 m3) of white pine; 328,000 cubic feet of red oak; 249,000 (6,902 m3) of soft maple; 175,000 of red pine (4,900 m3); 129,000 of quaking ash (3,612 m3), and 54,000 of white oak (1,512 m3). Dutch elm disease was probably why 22,000 elms were removed. Dutch elm disease is believed to have entered the United States in the late 1920s, and first found in Portage County in 1963. □

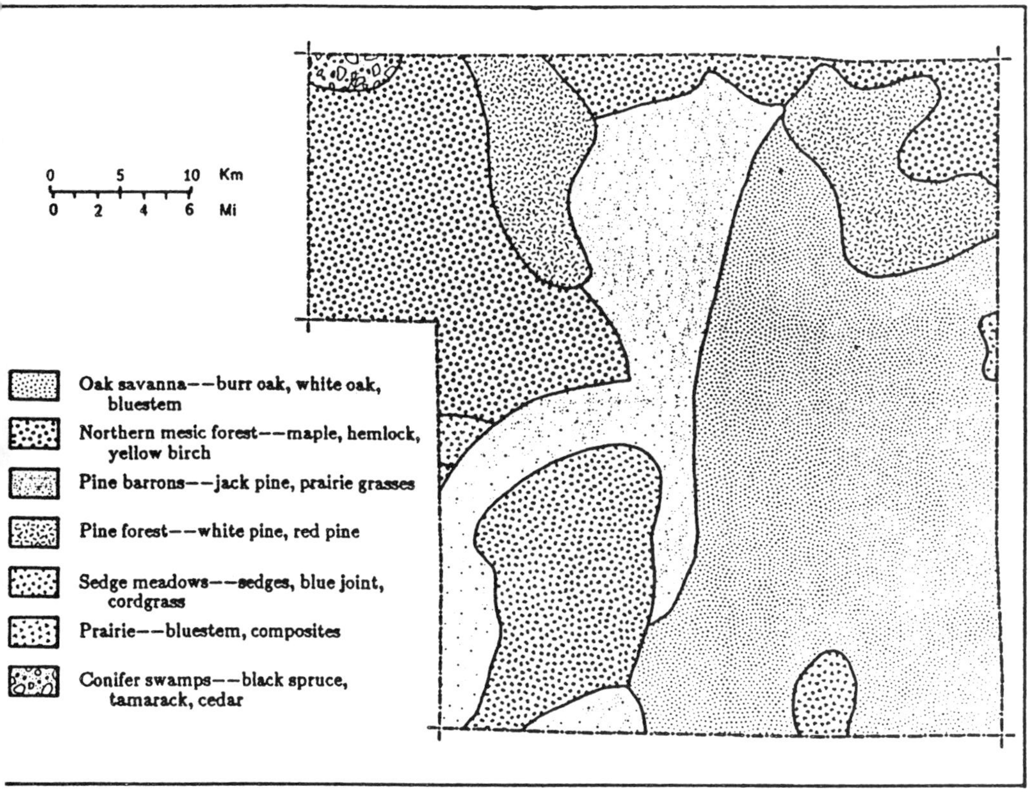

EARLY VEGETATION

Source: Surface Water Resources of Portage County
Dept. of Natural Resources, Madison Wis., 1972

RECREATION

With its variety of landscapes, including lakes, streams and marshes, forests and open land, hills and flat areas, forests and fields and farms and villages, Portage County offers all kinds of recreation.

Outdoor activities in summer start in April and end in November, peaking between Memorial Day and Labor Day. Many facilities are offered in the Stevens Point/Plover area and throughout the county for picnicking, camping, aquatic sports, fishing, bicycling, hiking, golfing and several other activities. These facilities consist principally of public parks: 18 county parks, 19 (of which 4 are not developed) in Stevens Point and the village of Plover has 15. Each of the other villages has at least one public park. There are 7 developed as town parks, and 33 designated as parkland but not developed yet. One state park, Hartman Creek State Park, has an area in Portage County, but the main area is in Waupaca county.

The most popular public parks in the county are:

<u>Bukolt Park, Stevens Point:</u>

Bukolt Park was originally an undeveloped picnic and playground located near the city's first waterworks plant, built in 1887. The area was designated as Waterworks Park. The city acquired the park in 1922. The waterworks ceased operation but the standpipe stood as a city landmark until its raising in 1956.

Great changes took place in the park in the 1930's: landscaping, a paved road, beautification of the lagoon, construction of a lodge and the establishment of a beach with a bathhouse. In 1936 Waterworks Park was renamed John J. Bukolt Park in honor of the founder of the Cradle Manufacturing Company, which became Lullabye Furniture Company. Since that time a picnic area, lighted baseball field, horseshoe pits, a boathouse, boat landing and fishing pier have been added.

<u>Goerke Park, Stevens Point:</u>

The present site of Goerke Park was first developed in the 1880's, after it was cleared of timber, and a race track was built. In 1885 the first annual Stevens Point fair was held there. And, in 1891, the fair was named the Great District Fair. Horse racing was the main attration. By 1927, however, interest in the trotters had waned and the last fair was held that year.

Two years later the city acquired the grounds from the county. The area was named for Mrs. Amanda Goerke, after her $10,000 donation in 1923. This gift was used to establish a trust fund for the development of the grounds as a recreational, school and military tract. It was then officially named "Goerke Memorial Recreational Field." Preparation for the use of the athletic field for track and football events was completed in 1932. It was during that fall that the field was first used for football games. By the fall of 1934, lights were installed and nighttime games were held. The first game took place on September 28, with Stevens Point defeating Wausau 27-7.

The field's stadium was completed in 1938 as a Works Progress Administration Project. The first building dedicated for athletic use was the Parker Memorial Fieldhouse. Work on that building began in 1933 through a $1,000 donation from Mrs. Laura Mae Corrigan in memory of her mother, Mrs. Emma Parker. Their construction was made with the help of federal funds and a work relief project. Construction materials were supplied by the city with much of them coming from dismantled buildings. Dedication took place on May 29, 1934 with music and athletic shows presented by local high school students.

In 1937 the largest structure on the grounds was constructed--P.J. Jacobs High School. Between 1937 and 1958 tennis courts, a public ice skating rink, a hockey rink and a municipal swimming pool were added to the field's recreational facilities. In addition, the field has a bathhouse, a concession stand, an indoor ice arena, a recreation center with television studio, gymnasium, meeting and game room and storage areas.

<u>Iverson Park, Stevens Point:</u>

Iverson Park was named after its donor, Jules Iverson. A Danish immigrant, Iverson came to Stevens Point in the 1870's and established a jewelry business. He later branched into real estate and mortgages and became a wealthy man. The park was donated to the city in 1935 with the understanding that it would be maintained in perpetuity as a public park and, especially, as a playground for children.

Federal Works Programs in the 1930's contributed greatly to the development of the park. In 1936 a community house, along with Boy and Girl Scout lodges, were built. Toilet and other sanitary facilities were added in 1938. A new swimming area was completed in 1938, and a new bathhouse was constructed during 1940. Ski enthusiasts were pleased by the erection of a ski scaffold in January 1937.

Since then other amenities have been provided including: a concession stand, lighted softball diamond, volleyball courts, picnic areas, nature trail and center, a sled hill, two toboggan and cross-country trails and a warming house.

Mead and Pfiffner Pioneer Park, Stevens Point:

Mead Park was donated by George W. Mead on behalf of Consolidated Power and Paper Company. The park included an area on both sides of the Wisconsin River. The first job in the creation of the park was the draining and filling of the areas that were below river level. Thereafter, seeding and shrubbery beds were constructed. In 1956 picnic tables were installed. Later, an open shelter was built along with two tennis courts, a volleyball court, a swimming beach and docking areas. The left bank became known as East Park which, in 1960, was changed to Pioneer Park because it was the site where the first settlers of Stevens Point landed. This name was later amended to Pfiffner Pioneer Park on account of a land donation on the north side, from the Pfiffner family. A bandshell was erected in 1976 on occasion of the bicentennial of the United States. The park has a lagoon, bathhouse, boat slip, fishing pier, lodge and picnic area.

In addition to these parks in Stevens Point, there are playgrounds in each of the public and parochial schools: five elementary and public, three parochial and one high school.

Collins Park:

Collins Lake was called Fish Lake from the 1870's until WWII. It was renamed Collins Lake after a country park was established on the north shore and named in honor of one time County Board chairman, W.F. Collins, of Arnott. The park has 25 campsites with utilities, picnic tables, beach, swimming raft, boat ramp, dressing area and shelter.

Jordan Park:

Jordan Park is located on both sides of Hwy 66 on the pond, which is an artificial pond created by the damming of the Plover river, and on the area below the dam. Fifty acres (20ha) are maintained as a natural area. There are 66 picnic tables, a beach on the pond, an enclosed shelter, 2 open shelters, a boat ramp, 25 campsites, a nature trail, a fitness trail, a nature center and a wildlife and waterfowl sanctuary. Below the dam is a canoe launch, which peolple use to float to Ivevrson Park. In the winter months there are cross sountry ski trails and snowmobile trails.

Lake Emily Park:

The history of Lake Emily Park goes back to the 1890's when the Old Settlers Club, organized in 1892 by members who settled in the county before 1860, had their annual picnic on the shore of the lake in 1894. As early as 1903 the County Board bought lands west of Amherst Junction to be used as the present Lake Emily Park. Later, the board created a Lake Park committee to develop a park. About 1910 the Lake Emily Park Committee started leasing lake shore lots from year to year to private individuals. Cottages then began to spring up in the north and east and, finally, on the south side in a little village around the lake--including the west end of the lake, which is privately owned. The largest amount of county-leased land on the lake took place in 1940 when 68 lots were rented out to private individuals. The leases, which all lot renters have signed, empowers the county to vacate part or all leased land, in order to provide more recreational space for the general public. Now, all leased land at the east end of the park has been discontinued and the cottage owners have removed their buildings. The lake offers facilities for swimming, motorboating, water-sking and fishing. The park also includes: 113 picnic tables, a beach, 2 enclosed shelters, a boat ramp, 19 campsites with utilities, restrooms, showers, playgrounds and a wildlife pen. Lake Emily is also the location of historical Indian Mounds.

Sunset Lake:

Sunset Lake is one of the most beautiful in central Wisconsin. It was first identified in the mid-1920's, when a portion of the east shore was acquired by the Boy Scouts, where it is now developed as Camp Chickagami ("Where camping is King"). The lake is completely in private hands. The public parkland is leased by the County Park commission from the New Hope Evangelical Lutheran Church, formerly the New Hope Norwegian Lutheran Church. It consists of a beach, a boat ramp, toilets and a bathhouse. It also has 25 picnic tables. The major uses are swimming, picnicking and fishing. Measured at 55 feet, it is one of the three deepest lakes in the county. (Budsberg Lake is 58 feet deep and Collins Lake is 56 feet deep. The next deepest lake is Lake Riley at 48 feet.) Although the figures published in 1990 are the same as those found in a report published in 1972, the lake has varied in elevation several times during these 18 years. The depth figures are only approximates.

The Dewey Shooting Range is also included in the County Park system. It was developed in 1986 and completed in 1987. The range was developed to provide a safe facility for sighting in guns and target practice. It is located in the Dewey Marsh in the Town of Sharon.

A trail linking the scenic natural areas of the Stevens Point/Plover area is named the Green Circle. Throughout the year it will accommodate walkers, hikers, joggers, bikers, photographers and cross-country skiers. Portions of the trail are presently in use, particularly along the Wisconsin River from Pfiffner Park to Bukolt Park.

In addition to the public parks and trails supervised by the government, there are streams and lakes which are used for boating, canoeing, fishing and swimming. The Wisconsin River is the principal one; dams have been erected and provide flowages which, in fact, are lakes.

The first dam goes back to 1846, when Abraham Brawley was authorized to construct it, but not more than six feet. In 1905 the Jackson Milling Company owned the dam site, which consisted of an old rock crib dam that was seven feet high and 800 feet long. The dam was located about 1500 feet south of the Clark Street bridge. The owner installed three tribunes, rated at 140 horsepower, to operated a flour and feed mill. In 1918 the Railroad Commission (later renamed Public Service Commission of Wisconsin) granted the owners a permit to raise, enlarge and rebuild the dam another 1,000 feet south of the original structure. Since that time the structure has seen the addition of a power house to furnish electricity for the Consolidated Paper Mills. In addition to the public parks, there are several routes for bicycles.

There are also a number of institutional facilities or privately owned facilities, many of which may be used by the public. They consist of a baseball diamond, a softball field, a quarter mile track, a hockey field, 12 tennis courts, 5 volleyball courts, 3 horseshoe pits, a football practice field, a rugby field and 4 flag football fields. Their usage is primarily for classes, secondly for University teams, clubs and organizations and, finally, for the public. Due to the priority given to campus use, University facilities are listed as semi-public. Schmeekle Reserve, supported by the University and open to the public, is 196 acres in size. It has a man-made lake covering 25 acres and provides visitors an enjoyable, "wild" community within the city of Stevens Point. In addition to its scientific use, trails provide opportunities for jogging, walking, biking and exercising.

The County Board, in 1935, created the County Park Commission. The purpose of this commission is the health, comfort, enjoyment and welfare of the people of the county. They also protect and preserve the lakes, streams and pools from pollution, as well as places of interest and historical sites. This commission publishes a comprehenseive outdoor recreation plan. The latest, a book of 127 pages printed in 1990, covers the period of 1990-1995. Today the commission, in addition to the population of the county, works with the Stevens Point/Plover Area Chamber of Commerce to attract tourists and conventioneers to Portage County.

GLOSSARY

ALLUVIAL PLAIN: A plain pertaining to, or consisting of alluvium; deposited from flowing water or pertaining to such a deposit.

AURIST: A specialist in regards to diseases of the ear.

BASALTIC: Of or pertaining to a kind of trap rock; a greenish or brownish-black rock, igneous in origin, of compact texture and considerable hardness, composed of augite or hornblende.

CARBON METHOD: The dating of materials through examining the long-lived radioactive isotope of carbon: carbon 14.

CUESTA: A landform commonly found in a region of gently tilted sedimentary rocks and consisting of an inclined upland slope, which conforms with the dip of a resistant bed and a relatively steep embankment descending abruptly from the crest.

DRESBACH FORMATION: A layer of sedimentary rock; e.g. sediment of sandstone of Cambrian age.

DRIFT: A term applied to any superficial deposit caused by a current of water or air; also Pleistocene deposits of glacial and fluvial-glacial detrimus, also known as boulderclay, and till.

DRUMLIN: A long narrow ridge of drift.

EOLIAN DEPOSITS: Deposits moved and left by the wind.

EPOCH: Any reference point in time.

ERRATICS: Stray masses of rock, foreign to the surrounding strata, and apparently brought from its original site by glacial action.

FAUNA: A collective term applied to animals or animal life of any particular region or epoch.

GEOPHYSICS: The science or study of the physics of the earth; especially its crust.

GEOSYNCLINE: A large depression in the surface of the earth, from the lowest point of which, there is a rise to either side of the depression.

GNEISS: A metamorphic rock composed like granite, of quartz, feldspar or orthoclase and mica, but distinguished from it by its foliated or laminated surface.

IGNEOUS ROCK: Rock resulting from or produced by volcanic agency.

KETTLE: A deep circular hollow scoured out in a rocky river bed or by a glacier.

LOAMY: Formed of earth.

LOBE: A deposit of fine yellowish-gray loam; considered to be composed of material transported by the wind during or after the glacial period.

MONADNOCK: A hill or mountain of erosion-resistant rock rising above a peneplain.

MORAINE: An accumulation of debris from the mountains carried down and deposited by a glacier.

METAVOLCANIC ROCK: Rocks resulting from volcanic activity.

OSTEOPATH: A specialist in regards to bones and skeletal structure. One who practices the method of cure founded on the assumption that the deformation of the skeleton and consequent interference with adjacent nerves and blood vessels are the cause of most diseases.

OUTWASH PLAIN: A plain beyond the terminal moraine where glacial material is deposited by glacial melt, usually composed of sand, gravel and, farther away, silt and clay.

PENEPLAIN: An erosion surface of considerable area and slight relief.

PLATTED: A section of land mapped in a plat book.

PLANIMETRIC MAP: A map without indication of elevation.

RESIDUUM: That which remains after a process of combustion, evaporation, etc.; a deposit of sediment; a waste or residual product.

RHYOLITIC: Referring to volcanic rocks which exhibit a fluidal nature.

SCHIST: A crystalline rock whose component minerals are arranged in a more or less parallel manner.

SIAL: This discontinuous upper layer of the earth's crust is represented by the continental masses which are composed of relatively light rocks rich in SILICA and ALUMINA and maybe regarded as flowing on a lower crystal layer of sima.

SIMA: The continuous basal layer of the earth's crust, made up of relatively heavy, basic rocks composed of SILICA and MAGNESIA, that underlies the sialic congenital masses and forms the crust under the oceans.

STRATUM: A layer of sedimentary rock with definite characteristics.

TECTONIC FORCES: A term applied to the motion associated with continental movement.

BIBLIOGRAPHY

Hundreds of books have been consulted by students and the author on Portage County, Wisconsin and the topics treated in the present book. The readers who are interested in more details should consult the original student papers at the Library of the Portage County Historical Society in the Learning Resources building of the University of Wisconsin-Stevens Point.

The most important works are the following:

Gard, Robert L. and L.G. Sorden: *The Romance of Wisconsin Place Names*, October House Inc. New York 10011, 1968.

Holt, C.L.R. Jr.: *Geolgy and Water Resources of Portage County*, Wisconsin, 1965.

Martin, Lawrence: *The Physical Geography of Wisconsin*, second edition, Wisconsin, 1932.

Portage County Comprehensive Outdoor Recreation Plan 1990-1995 Update, Portage County Parks Commission, Stevens Point, 1990.

Soil Survey of Portage County, Wisconsin, United States Department of Agriculture Spill Conservation Service, 1978.

Stevens Point Centennial, 1858-1958, Portage County Historical Society, 1958.

Surface Water Resources of Portage County, Department of Natural Resources, Madison, Wisconsin, 1972.

Maps

U.S. Geological Survey:

Scale
1:48,000 (15 minutes) (planimetric)
1:24,000 (7.5 minutes) (topographic)
Eau Pleine, Dancy, Dewy Marsh, Rosholt W, Rosholt Sherry, Junction City, Rocky Run, Stevens Point, Polonia, New Hope, Tigerton, Meehan, Whiting, Arnott, Amherst, Scandinavia, Kellner, Bancroft, Almond, Blaine, King.

Atlas & Plat Book, Portage County, Wisconsin (various editions), Rockford Map Publishers, Inc. Rockford Illinois (scale 1.5 inch to the mile).

LIST OF STUDENT STUDIES
ON
PORTAGE COUNTY

PHYSICAL GEOGRAPHY

Geology and Landforms

Physical Geography: Tom Bellile	1970
Geology: Steven Kahl	1972
Geology: Eric W. Nilsson	1969
Geologic Formations: Paul Ditter	1970
Precambrian Boundary: Ray Hartl	1972
Precambrian Rock Formations: Steven Smith	1970
Outliers: Jackson P. Grode	1969
Glaciation in the "Driftless Area": James D. Janke	1970
Why Part of Portage County was Never Glaciated: Mike Gallenberger	1969
Terminal Moraine Region: Pete Jungenberg	1971
Arnott Moraine: Kenneth Rierson	1968
Outwash Plain: John Valiga	1979
Kettles: Christopher Young	1974
Drumlins and Eskers: Winifred Anne Ludtke	1979
Heavy Mineral Survey: Robert Cook	1977
Continental Divide: Kent Wilson	1968
Continental Divide: Roger Messar	1977
Sand Dunes: Dennis Weitzman	1972
Terraces of Iverson Park: Walter J. Mockler	1972
Landform Classification: Phillip Manske	1979
Landforms and Land Use Relationships: David Hogan	1981
Landforms in the Town of Almond: Tom Mickewicz	1970
Landforms in the Town of Buena Vista: James F. Runke	1970
Landforms in the Town of Linwood: Mike Majewski	1974
Landforms in the Town of Lanark: John Davis	1968
Landforms and Land Use in the Town of New Hope: Samuel E. Smith	1981
Landforms in the Town of Sharon: John Price	1972
Regions: Cheryl Hass	1969
Stone (under Cultural Geography: Settlements and Architecture): Catherine Johnson	1970

Climate

Climate Social and Economic Effects: Gregory J. Hanson	1976
Climatic Statistical Data for Coddington: Harris Moskonas	1967

Hydrography

Water: Stephen W. Justinger	1978
Drainage: John Rebholz	1973
Wisconsin River: Bruce Wegner	1970
Wisconsin River Survey Project: John L. Koziol	1970
Wisconsin River Source of Pollution: J.M. Labiszak	1970
Flooding of the Wisconsin River in the Stevens Point Area: George Gitter	1981
Plover River: John D. Cory	1975
Little Plover River: Ken Jagodzinski	1971
Tomorrow River: Dave Wood	1971
Mill Creek: Doug Shapyak	1972
Poncho Creek: Douglas A. Severson	1972
Love Creek: Steve Kowalkowski	1979
Lakes: Mary B. Cychosz	1970
Lakes: Robert McGrane	1968
The Origin of Lakes: Erik Tolvstad	1973
Fluctuation of Lake Levels: Eugene C. Cebulski	1980
Lake Levels: Ronald Dudley	1978
Wetlands in the Towns of Hull and Dewey: Lucy A.K. Oyebog	1978
Buena Vista Marsh: Greg Waterhouse	1975
Drainage Ditches: Richard Lahaie	1975

Vegetation

Tree and Soil Relations in the Town of Hull: Peter Konopacky	1971
Tree Species in the Town of Linwood: Chris Velden	1979
Trees in the Town of New Hope: Christine Perszyk	1976
Tree Species in the Town of Sharon: Richard A. Runquist	1971
Tree Species in the Town of Stockton: Donald Denor	1972
Elm and Dutch Elm Disease: Jack Westphal	1972

Cultural Geography

History

Archaeological Sites: Herbert Peterson	1976
Historical Map: Keith Karow	1976

Population

European Settlement in Wisconsin 1870-1960: William Yanacheck	1973
Ethinic Groups and Farmsteads: Paul J. Crerar	1976
German Influence: Gene Gohlke	1968

Irish Inhabitants of Portage County: Partick O'Leary	1973
Scandinavian Settlement: Chuck Fallos	1970
Scandinavian Element: Allan Christianson	1972
Individual Isolated Scandinavian Settlement: Ronald Wage	1970
Polish Element: Bob Kindler	1971
Polish Elements: Thomas E. Peckham	1981
Amish: Rodney Anderson	1970
Amish: Tori Boers	1981
Our Amish Neighbors: Patrick Huempfner	1977
Mexican-Americans in Portage County: John O'Dell	1975

Cultural

Religion and Cemeteries

Religions: Kriang Kiatfuengfoo	1969
Catholic Churches: Mike Counsell	1969
Lutheran Churches: Thomas R. Lubner	1971
Churches in Stevens Point: Fred Melicher	1975
Rural Cemeteries: James Wieser	1971
Crosses and Shrines: William Schultz	1974

Schools and Education

Schools: Quinn Vanden Heuvel	1970
Territoriality of Portage County Schools: Wayne C. Peterson	1974
Stevens Point Public Schools: Daniel A. Nowak	1971
Public Elementary Schools: Harvey J. Eckert	1972
Consolidation of Rosholt School District: John J. Wanserski	1975
History of WSU-SP: Ed Terzynski	1970
Campus Facilities: Jim Donlon	1975
Housing and Educational Expansion: Thomas F. Behm	1969
Student Housing in Stevens Point: John D. Davies	1972
UWSP Commuter Student: Michael V. Hannum	1975
The Present Use of Rural Schoolhouses: Randy Miller	1973

Settlements and Architecture

Horizontal Log Timber Structures: Ned Braatz	1969
Stone Structures: Catherine Johnson	1970
Barns: Ned Braatz	1969
Early Architecture of Stevens Point: Ronald G. Kappell	1969
Domestic Architecture in Stevens Point: Eric Holm	1976
Brick Buildings in Stevens Point: W. Yanacheck	1972
Victorian Styles of Homes in Stevens Point: Tom Kinsella	1973
Church Architecture in Portage County: John Neu	1973

Apartment Complexes: Richard Jonely	1974
Trailer Courts: David Jelmeland	1971
Town Halls of Portage County: Randy Miller	1974

Urban Studies: Stevens Point

North Central: Alan Stalbaum	1970
North: Bill Kruger	1971
Northeast: Dan Peterson	1971
North West Central: Dan Leider	1968
South Side: Bill Russell	1969
South West: Jim Sippel	1971
South West Central: John Holman	1972
West Central: James Schultz	1972
Public Square Area: Jerry Ross	1971
Public Square: Gary Van Rixel	1981
CBD of Stevens Point: Diane C. Fiedler	1973
Annexation from the Town of Hull: Steve Catlin	1968
Annexation of Land in Stevens Point: Dennis Cook	1972
Comparison of Land Use and Zoning in Stevens Point: David Regan	1974

Urban Studies: Places outside Stevens Point

Almond: Kurt Dikkers	1971
Amherst: Charles Miescke	1970
Amherst Junction: Michael J. Hovel	1972
Amherst Junction: Troy Burns	1982
Arnott: Jeanne M. Rorge	1977
Bancroft: Terry Bauer	1971
Coddington: Harris Moskonas	1969
Custer: Rodney Hinz	1971
Ellis: Richard Wilcox	1969
Junction City: Thomas F. DeLoughery	1970
Kellner: John Abler	1972
Nelsonville: Stephen Meyer	1968
Nelsonville: Jayne Goehler	1984
Park Ridge: Mark Farah	1977
Plover: John Norton	1968
Plover: Dennis Shaw	1981
Polonia: Brian Burkhart	1971
Rosholt: James Collins	1971
Whiting: Wayne Kurtzweil	1972
Unincorporated Villages: Larry Olenchek	1972
Functions in Six Villages: Almond, Amherst, Amherst Junction, Junction City, Nelsonville, Rosholt: Jeff Darga & Scott Prefontaine	1983

Economic Geography

Agriculture

Barns (under Cultural Geography: see Settlements and Architecture; includes History of Farming, Fences): Ned Braatz	1969
Study of Two Farms located in Adjoining sections: Mark N. Smolen	1971
A Norwegian Farm: David Ninneman	1981
Keen Family Farm 1885-1980: Debbie Kluz	1980
Dairy Distribution: Mark Hinterberg	1979
Present Day Dairy Farming: Linda Literski	1980
Dairying in the Town of Belmont: Gerald A. Disterhaft	1973
Dairying in the Town of Grant: Gary Cook	1974
Dairy Farms in the Town of Hull: John Wright	1973
Dairy Farms in the Town of Linwood: Larry L. Koy	1972
Dairying in the Town of Stockton: Ron Ledvina	1968
Dairy Farming in the Town of Stockton: Wayne A. Binning	1971
Corn Production: John Hopkins	1981
Potato Growing: Bruce L. Decker	1969
Potato Production: Jeffrey B. Stroub	1973
Cranberry Growing: Greg Wing	1971
Cranberry Industry: George E. Sass	1972
Snap Beans, Peas, and Cucumbers for Canning: Roger Anderson	1979
Christmas Tree Growing: A specialized Horticulture: John Lagerman	1977
Game Bird and Animal Farms: Charles Payant	1975
Irrigation: John Cychosz	1973
Fences: Dennis LaFratta	1972
Rural Fences: Craig Varland	1976
Cropland Adjustment Program: Richard Korpela	1972
Buena Vista Marsh Economics as Related to Soil Types: Douglas C. Knox	1968
Abandoned Farms of the Town of Almond: Jan Greenquist	1971
Abandoned Farms in the Town of Dewey: Kiane Kokaly	1976
Abandoned Farm Houses in the Town of Lanark: Mark Barber	1972
Abandoned Farms in the Town of New Hope: William Rindal	1970
Abandoned Farms in the Town of Sharon: David R. Briggs	1972
Abandoned Farms in the Town of Stockton: Dennis A. Rasmussen	1968

Industry

Industries in Portage County: Robert Simkins	1968
Stevens Point Area Manufacturers: Carrie Papa	1981
Wisconsin River Industries: Bill Freude	1980
Logging and Lumbering: Laurinda Haas	1980
White Pine Industry: Brian C. Johnston	1971
Paper Industries in Portage County: Mike Lynott	1975
Stevens Point Division of Consolidated Papers, Inc.: Richard A. Hartman	1970

Wisconsin Division of Consolidated Papers, Inc.: Rita Rae Dins 1968
Paper Mill Dams: James Washburn 1970
Stone (Under Cultural Geography: Settlements and Architecture):
Catherine Johnson 1970
Sand and Gravel Industry: Terry Hegeman 1970
Trucking Industry in Stevens Point: Ronald Carlson 1972
Industrial Park and Industrial Development of Stevens Point: Charles L. Gust 1974
Stevens Point Industrial Park: Mark Buschman 1980

Transportation and Traffic

Transportation System: Thomas Bright 1968
Railroads: Bruce Berntzen 1970
Railroads and the Development of Portage County: Robert Rieckmann 1970
Railroad Facility Characteristics of Selected Towns and Village in
Portage County: Bradd W. Price 1971
Stevens Point and the Railroad: Le Roy Gura 1970
Economic Impact of the Removal of Railroad Lines within Portage County:
Robert C. Wundrock 1972
Effects of the Chicago and North Western's Proposed Abandonment of its
Ripon to Bancroft Line on Portage County: Jim Mielke 1980
Air Transportation for Portage County: David Plamann 1970
Federal Aviation Administration in Central Wisconsin: Franklin Schumacher 1974
Report on Airports of the Area: Stevens Point Municipal Airport, Wausau
Municipal Airport, Marshfield Airport: Greg Puza 1973
Stevens Point Municipal Airport: Lawrence Goeb 1972
Indian Trails versus Existing Transportation Trails: Brian Ratzburg 1980
Highway Study: Gary M. Nelson 1972
Roads: David E. Christensen 1970
Bridges: John Salmon 1973
Public Transportation in Stevens Point: William DeJung 1979
Bikes: Scott C. Gilmore 1974
Meter-controlled Parking Facilities in Stevens Point: William H. Monger 1973
Automobile Parking at Wisconsin State University Stevens Point:
Dennis A. Ahonen 1971
Parking Facilities of the University of Wisconsin Stevens Point:
Mary Brumm 1977
Computer Workers of the Stevens Point Area: Edward Hertel 1973
UWSP Commuter Student (under Cultural Geography: Education):
Michael V. Hannum 1975
Traffic Survey in Stevens Point (Main Street Traffic): Lynn R. Dhein 1969
Traffic Survey in Stevens Point (Junction Union Street and Maria Drive):
Donald Zander 1970
Accident Survey of Stevens Point: Mark Mentch 1975
Traffic and Accidental Survey in Stevens Point: Dudley Zhe 1969
Traffic Plan for Stevens Point (Main and Clark Streets): Donald Pochowski 1973

Rerouting Highway 10: Bruce Bullard	1980

Trade and Services

Stevens Point Retail Service Area: Raymond Madsen	1973
Shopping Centers in Stevens Point: Steve Waters	1978
DownTown Mall Project in Stevens Point: Margaret Laatsch	1983
Distribution of Grocery Stores in Stevens Point: Daniel Marx	1968
Grocery Stores in Stevens Point: Barbara Rudolph	1973
Portage County Grocery Stores (Excluding Stevens Point Area): Burt D. Thorp	1974
Grocery Stores in Stevens Point Area: Bill McCartney	1975
Bakeries in Portage County: John Peotter	1975
Sausage Making and Meat Processing: Lee Smolen	1974
Drug Stores of Portage County: Michael T. Tlapa	1974
Hardware Stores in Stevens Point: Robert J. Kellett	1980
Portage County Liqour Stores: Scott Rutter	1978
Wholesale Beer Distribution in Portage County: John Woehler	1974
Distribution of Portage County Bars: Jim Burns	1972
Taverns on the Public Square in Stevens Point: Ron Bornhauser	1975
Restaurant Industry of Stevens Point: Mary Wanee	1970
Restaurant Row in Stevens Point: David Plaisance	1983
Motels and Hotels in the Stevens Point Area: Thomas Mangene	1973
Hotels and Motels of Stevens Point: Paul Boyer	1975
Service Stations in the Stevens Point Area: Richard Hartman	1972
Gasoline Stations of Stevens Point: Gregory A. Kobs	1975
New and Used Car Dealers in Portage County: Harold L. Bolstad	1975
Auto Salvage Yards in Portage County: James C. Wollerman	1974
Motorcycle Sales and Service in Portage, Marathon and Wood Counties: David R. Birge	1974
Bikes (Under Economic Geography: Transportation and Traffic): Scott C. Gilmore	1974
Billboards as Advertising: Rick Baumgart	1975
Banks of Portage County: Raymond Madsen	1973
Financial Institutions: Richard Kellogg	1979
Financial Institutions: Michael J. Franks	1984
Health Care Facilities Within the City of Stevens Point: John C. Sturdy	1977
Funeral Homes: Thomas J. Stobbs	1973
Telephone Companies: Jim Gluth	1974
Sounds in the Air -- Radio Stations: William Pelke	1973
Portage County Sheriff's Department: Gary Mueller	1979
Water Use: Robert Guse	1975
Electricity: Dale Larkee	1984
Supply and Transport of Natural Gas: Tom Braun	1970
Pipeline Operation in Stevens Point: Paul Ehlers	1974
Sewage Treatment: Bruce Moebius	1978
Dumps of Portage County: Mark Koenig	1979

Stevens Point Area Garbage Renewal and Current Landfill Development: Charles M. Fondrie	1981
Portage County Fire Departments: Don Behnke	1975
Fire Control in Portage County by Volunteer Fire Departments and the Department of Natural Resources: James LaMar	1973
Portage County Weather Stations: Tim Koenig	1974
Military Detachments in Stevens Point: Keith W. Bowles	1975

Recreation

Outdoor Recreation: Neil Dreifurt	1976
Outdoor Recreation, Private/Commercial: Thomas E. Tews	1980
Outdoor Recreation in the Stevens Point Area: Bruce Klocke	1970
Economic Aspect of Summer Recreation: Garry Therriault	1971
Outdoor Winter Recreation in Wisconsin: Tom R. Thielke	1968
Portage County Parks: John Schumacher	1972
Stevens Point Park System -- From the Past into the Future: Glen Pelo	1974
State Owned Public Hunting and Fishing Grounds: Kurt Bernhagen	1977
Hunting: Brian K. Fay	1978
Fishing: Paul J. Schneider	1977
Trapping, A Modern Look at an Ancient Art: Richard W. Rudis	1980

Planning

Land Use Planning: Morris Wilhelm	1981
DownTown Mall Project in Stevens Point (under Economic Geography: Trade and Services): Margaret Laatsch	1983
Traffic Plan for Stevens Point (Main and Clark Streets under Economic Geography, Transportation and Traffic): Donald Pochowski	1973
Stevens Point Park System -- From the Past into the Future (under Economic Geography, Recreation): Glen Pelo	1974

LAND USE OF SECTIONS AND SPECIAL AREAS

Township North	Range East	Section	Author	Year
21	8	23	Patricia Schwerin Murphy	1970
22	7	29	Douglas Bathke	1971
22	8	1	John D. Fullmer	1968
22	9	13	Guy Stanoch	1978
22	10	9	Joan C. Meyer	1974
22	9	8	Michael Schroeder	1975
23	7	30	Jim Reimer	1976

Township North	Range East	Section	Author	Year
23	8	3	Robert C. Domagalski	1970
23	8	13	Tom R. Thielke	1968
23	8	14	Kenneth Petersen	1968
23	8	16	Larry Johnson	1971
23	8	26	Gale West	1968
23	8	26	John Roselius	1977
23	8	36	Janice Kooiman	1980
23	9	18	William H. Reichwald	1970
23	9	24	Jeff Pierce	1971
23	9	26	Gene M. Wanta	1971
23	9	34	Eric K. Smedal	1981
23	10	12	Jim Avila	1979
23	10	20	Robert Levknecht	1977
23	10	28	John P. Schmelling	1971
24	6	30	John Pupols	1970
24	7	6	William J. Werner	1975
24	7	21	Donald Hildebrandt	1973
24	7	29	Emery W. Henrich	1969
24	8	8	Tom Ebel	1973
24	8	9	Tim Muench	1972
24	8	12	Ken Nowak	1976
24	8	14	William R. Vachon	1979
24	8	19	Mike DeLoughery	1972
24	8	20	Elmer Siegel	1970
24	8	21	Richard Newgord	1974
24	8	21	Ed Terzynski	1970
24	8	21	Thomas C. Jauquet	1974
24	8	23	Bob Kettner	1973
24	8	26	Craig Hansen	1968
24	8	27	Walter J. Mockler	1972
24	8	32	Curt Conrad	1973
24	9	7	Larry Banaszynski	1972
24	9	16	August J. Reichardt	1978
24	9	24	Dean L. Baggenstoss	1978
24	9	29	Jim Pierson	1970
24	9	31	Don Hinz	1971
24	9	32	Donald Dilges	1974
24	10	19	Larry E. Shadick	1971
24	10	22	John Burbey	1971
24	10	24	Mark Hillegas	1971
24	10	30	Immanuel Johnson	1974
24	10	34	John V. Olson	1972
25	6	31	Steve Sheets	1978
25	6	36	Randy Hendrickson	1975

Township North	Range East	Section	Author	Year
25	7	19	David B. Grueschow	1971
25	7	20	Rick DeFauw	1970
25	8	13	Edward Heil	1971
25	8	33	Steven Georgeff	1976
25	8	36	Richard Klein	1968
25	9	7	John Auman	1977
25	9	16	Kenneth P. Boyd	1978
25	10	16	Pat Duffek	1971
25	10	22	Jack Prahl	1971

Shoreline Development of Lake DuBay in Portage County:	Jim Francken	1973
Hilladay Creek Bottomland:	Robert Bochenek	1970

Cornerstone Press

Each fall Cornerstone Press springs to life as University of Wisconsin - Stevens Point Editing and Publishing students accept positions in the company. Supervised by Dan Dieterich, Chief Executive Officer and teacher, students push at least one manuscript through the entire publication process. From selection to editing, design to marketing, the class is involved in every aspect. Cornerstone Press works as an actual publishing firm, soliciting the university for funds, signing contracts with authors and making arrangements with printers and book stores. It is a unique opportunity for interested students to gain hands on experience in the publishing community.

Chief Executive Officer: Dan Dieterich
President: Stephanie Raymaker
Editor in Chief: Thaddeus J. Brockman
Production Manager: David S. Dortman
Business Manager: Julie A. Block
Corporate Secretary: Lila Gallenberg
Fulfillment Manager: Jeff Scanlan
Managing Editor: Jo Ellen M. Maxymek
Substance Editor: Pamela J. P. Schroeder
Assistant Substance Editor: Gretchen Rybicke
Copy Editor: Wendy Robinson
Assistant Copy Editors: Brenda Pagel, B.J. Hiorns
Assistant Production Manager: Julie Postier
Production Assistant: Ann M. Wege
Designer: Valentina Kaquatosh
Subsidiary Rights Manager: Joe Ruelle
Marketing Manager: Paul Matsuda
Sales Manager: Nina Pelo
Advertising Manager: Peggy Zabrowski
Publicity Director: Terrie Ann Shipley
Assistant Publicity Director: Lisa Lemke